ADDITIVE MANUFACTURING
Foundation Knowledge
for the Beginners

T0321951

Other World Scientific Titles by the Author

An Introduction to Electrospinning and Nanofibers
ISBN: 978-981-256-415-3
ISBN: 978-981-256-454-2 (pbk)

An Introduction to Biocomposites
ISBN: 978-1-86094-425-3
ISBN: 978-1-86094-426-0 (pbk)

Polymer Membranes in Biotechnology:
Preparation, Functionalization and Application
ISBN: 978-1-84816-379-9
ISBN: 978-1-84816-380-5 (pbk)

The Changing Face of Innovation: Is it Shifting to Asia?
ISBN: 978-981-4291-58-3 (pbk)

ADDITIVE MANUFACTURING
Foundation Knowledge for the Beginners

Sunpreet Singh
Chander Prakash
Seeram Ramakrishna

National University of Singapore, Singapore

 **World Scientific**

NEW JERSEY · LONDON · SINGAPORE · BEIJING · SHANGHAI · HONG KONG · TAIPEI · CHENNAI · TOKYO

Published by

World Scientific Publishing Co. Pte. Ltd.

5 Toh Tuck Link, Singapore 596224

USA office: 27 Warren Street, Suite 401-402, Hackensack, NJ 07601

UK office: 57 Shelton Street, Covent Garden, London WC2H 9HE

Library of Congress Cataloging-in-Publication Data

Names: Singh, Sunpreet, 1989– author. | Prakash, Chander, author. | Ramakrishna, Seeram, author.
Title: Additive manufacturing : foundation knowledge for the beginners / Sunpreet Singh,
    Chander Prakash, Seeram Ramakrishna, National University of Singapore, Singapore.
Description: Hackensack, NJ : World Scientific, 2021. |
    Includes bibliographical references and index.
Identifiers: LCCN 2020027898 | ISBN 9789811224812 (hardcover) |
    ISBN 9789811226243 (paperback) | ISBN 9789811224829 (ebook) |
    ISBN 9789811224836 (ebook other)
Subjects: LCSH: Additive manufacturing.
Classification: LCC TS183.25 .S56 2021 | DDC 621.9/88--dc23
LC record available at https://lccn.loc.gov/2020027898

British Library Cataloguing-in-Publication Data
A catalogue record for this book is available from the British Library.

For any available supplementary material, please visit
https://www.worldscientific.com/worldscibooks/10.1142/11953#t=suppl

Desk Editor: Shaun Tan Yi Jie

Typeset by Stallion Press
Email: enquiries@stallionpress.com

Printed in Singapore

Preface

Additive Manufacturing is one of the hottest topics in today's educational, research, and industrial environments. Owing to this, numerous educational institutes and universities across the world are now preparing their graduates to face the industrial demands of Additive Manufacturing technologies. However, it has often been observed that young students are unable to understand the basic fundamental principles, work procedures, type of suitable feedstock systems, and emerging applications of Additive Manufacturing. Despite the availability of a wide range of textbooks, reference materials, and literature information, the lack of a fundamental book that can help young students inculcate the foundation knowledge still exists.

Therefore, the present book has been designed and structured to provide the technical insights of Additive Manufacturing technology in a way that overcomes the learning barriers faced by young students. Particularly, the text has incorporated reading material to suit the curriculum requirements of undergraduate and graduate students across the world. The authors have classified the entire content into six focused chapters. Apart from discussing the basics of Additive Manufacturing technology in the first four chapters, chapters 5 and 6 focus on special purpose Additive Manufacturing technologies and testing and measurement of manufactured products. We are highly certain that the book will address the learning constraints of students and suit their academic requirements.

Sunpreet Singh
Chander Prakesh
Seeram Ramakrishna

About the Authors

 **Dr Sunpreet Singh** is a Research Fellow at the Department of Mechanical Engineering, National University of Singapore, working on the 3D printing of high-performance thermoplastics. He has developed various polymeric feedstock materials for specialized engineering and biomedical applications. During his extensive research experience with 3D printing technology, he has contributed more than 100 research articles in various reputed journals including Springer, Emerald, Elsevier, Wiley, and others. He has filed several patents and edited more than 10 scientific books, including *Biomanufacturing* (Springer, 2019), *3D Printing in Biomedical Engineering* (Springer, 2020), and *Functional Materials and Advanced Manufacturing* (CRC Press, 2020). He is the guest editor of several journals. During his teaching tenure, he has taught "Reverse Engineering and 3D Printing" and "3D Printing Technology" to undergraduate students. He received the Research Excellence Award for the year 2017–2018 from Lovely Professional University, India.

Dr Chander Prakash is Associate Professor and Head of the Industrial Engineering Department in the School of Mechanical Engineering at Lovely Professional University, India. He is a lifetime member of the Institution of Engineering and Technology (IET) and the Indian Science Congress Association (ISCA). He holds 13 patents, and has published more than 100 articles in peer-reviewed international journals and conference proceedings. He is also the guest editor of 10 books, including *Biomanufacturing* (Springer, 2019), *3D Printing in Biomedical Engineering* (Springer, 2020), and *Functional Materials and Advanced Manufacturing* (CRC Press, 2020). Dr Prakash's research concentrates on the synthesis/development, surface modification, and advanced/precision machining of metallic and non-metallic biomaterials. He explores three main research directions: synthesis/development of magnesium/titanium-based biodegradable alloys and composites, surface modification of biodegradable polymeric and magnesium/titanium-based biomaterials, and precise and advanced machining of biomaterials using electric discharge machining, magnetic abrasive finishing, and diamond turning processes.

Professor Seeram Ramakrishna, *FREng* is the Director of Center for Nanofibers and Nanotechnology at the National University of Singapore (https://www.nature.com/articles/nj0232.pdf?draft=marketing). He is regarded as the guru of electrospinning and nanofibers (http://nart2020.com/conferenceinfo/). Microsoft Academic ranked him among the top 36 salient authors out of three million materials researchers worldwide (https://academic.microsoft.com/authors/192562407). He is named among the World's Most Influential Minds (Thomson Reuters), and the Top 1% Highly Cited Researchers in materials science and cross-field categories (Clarivate Analytics). He received his PhD from University of Cambridge, UK; and The TGMP from Harvard University, USA. He is appointed as the Honorary Everest Chair of MBUST, Nepal. He is an elected Fellow of UK Royal Academy of

Engineering (*FREng*); Singapore Academy of Engineering; Indian National Academy of Engineering; ASEAN Academy of Engineering & Technology; International Union of Societies of Biomaterials Science and Engineering (FBSE); Institution of Engineers Singapore; ISTE, India; Institution of Mechanical Engineers and Institute of Materials, Minerals & Mining, UK; American Association of the Advancement of Science; ASM International; American Society for Mechanical Engineers; American Institute for Medical & Biological Engineering, USA; and International Association of Advanced Materials (FIAAM). He chairs the Circular Economy taskforce at NUS, and is a member of Enterprise Singapore's and International Standards Organization's Committees on ISO/TC323 Circular Economy and Circularity. He is the Editor-in-Chief of Springer NATURE journal *Materials Circular Economy*. He is an editorial board member of Springer NATURE journal *Advanced Fiber Materials*; Elsevier Journal *Current Opinion in Biomedical Engineering*; and NATURE *Scientific Reports*. He is an opinion contributor to the Springer Nature *Sustainability Community* (https://sustainabilitycommunity. springernature.com/users/98825-seeram-ramakrishna/posts/looking-through-covid-19-lens-for-a-sustainable-new-modern-society).

Contents

List of Figures

List of Tables

Chapter 1

Introduction to Additive Manufacturing

This chapter introduces the reader to the multiple domains of Additive Manufacturing (AM) technology, including its history, principles, procedural steps, and classifications. This chapter guides young students in learning the basics of AM so as to enable them to understand information provided in the succeeding chapters.

1.1. History of AM

AM technology was earlier referred as Rapid Prototyping (RP), which first emerged in 1987 with the development of the first commercial system "Stereolithography" by 3D Systems Corporation, a process that solidifies thin layers of ultraviolet (UV) light-sensitive liquid polymer using a laser [1]. Since then, strong public interest has acted as a driving force in the advancements of this technology, leading to the development of several similar systems in a short period, while many more refined versions are still in progress. The earliest applications of this technology, as RP, were focused to provide an efficient platform for designing and validating the engineering concepts. Along the way, it has also overcome the potential challenges, including direct and indirect costs, tooling, and facilities required for maturing a conceptual prototype. Gradually, the RP technology has grown to offer the following benefits:

- *Enables the designers to explore, test, and validate the concepts:* It means that the efficient technological systems can produce a conceptual design with less time and cost.
- *Refine a concept:* RP helps in minimizing the design mistakes associated with small volume production runs.
- *Provides a more reliable platform for continuous breakthroughs:* The physical design models can be easily modified by applying repeated designs changes that allow for evaluation of the products.
- *Effective communication:* With the help of physical prototypes created via this technology, it is now easy for hands-on product experience amongst the designers, clients, and collaborators.
- *Optimum material acquisition:* These systems consume an optimum volume of the raw material and leave no scrap behind. Indeed, some systems utilize support materials which can be recycled easily.
- *Saves resources:* The independent RP systems do not require any type of additional tooling, therefore saving significant amounts of time and monetary resources for commercial enterprises.

The RP technologies have been continuously evolved, for more than a decade, and have become a vital segment of many industrial sectors. Apart from physical prototyping, some of the advanced applications of RP include functional prototypes, patterns for castings, medical models, artworks, and models for engineering analysis [2]. It is worth mentioning that the extensive use of RP has also led to the simultaneous development of application-specialized software tools and build materials. From 2000 to 2005, industries considered the RP technologies as a Rapid Manufacturing (RM) tool; therefore in 2009, the American Society for Testing and Materials (ASTM) and The International Organization for Standardization (ISO) defined RP as AM that is entirely opposite to the conventional subtractive manufacturing and formative manufacturing methodologies [3]. The term AM was formally adopted in 2009, by ASTM F42 Technical Committee members. The team had also proposed another name "three-dimensional (3D) printing" that was eventually not selected. Despite the instructions of the ASTM F42 Technical Committee, numerous terminologies including 3D printing, additive fabrication, additive processes, additive techniques, layer manufacturing, free-form fabrication, rapid tooling, additive layer manufacturing, solid free-formed fabrication, rapid prototyping, rapid manufacturing, and direct digital manufacturing are being used interchangeably by the scientific and engineering community.

1.2. Principle of AM

AM involves the fabrication of a product in the form of pre-defined slices of layers on a fixtureless platform. Being completely opposite to the conventional/subtractive manufacturing technologies whereby the process starts with a block of material and unwanted material is progressively removed until the desired part is obtained, the products developed by AM technologies follow an "Additive" principle. The principle of AM can be defined as:

"Progressive addition of thin slices of feedstock layers in a layer-upon-layer fashion"

This means that the raw feedstock material, in a pre-determined format, has been selectively deposited and, collectively, the process is referred to as "Additive Manufacturing". Although the basic principle of AM is not new, the way by which the material is processed using a digitally defined blueprint is unique. In AM systems, physical components are made from virtual computer-aided models. Furthermore, AM starts with nothing and builds a part one layer at a time by depositing each new layer on top of the previous one, until the part is complete. The layer thickness can range from a few microns up to around 0.25 mm, depending on the type of feedstock and AM technology used [4]. The unique principle and manufacturing ability of AM offer the following merits and demerits, when compared to conventional/subtractive manufacturing [5]:

In addition, with the use of AM, a computer-aided design (CAD) can be directly transformed to a finished/semi-finished product by eliminating the use of additional fixtures and cutting tools. Moreover, the AM

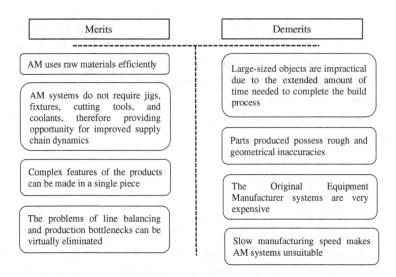

principle enables the production of complex and intricate products which generally are not feasible to manufacture using conventional machining operations. The sophisticated AM techniques also enable the fabrication of the hard-to-machine metals and alloys. In sum, the AM principle facilitates environmental friendliness, design flexibility, and lean production of the functional components. However, there exists some AM technologies, for example fused deposition modelling, that utilize a support material to facilitate the fabrication of overhanging sections of the products. The building of such support structures has also been executed following the "Additive" principle and the material used is generally different.

1.3. Procedure of AM

The AM procedure includes multiple stages, where the digital data is transformed from virtual prototype to solid physical object. However, each stage plays an important role in obtaining the desirable quality characteristics of the end-user product. The systematic procedural steps of AM technologies, as depicted in Figure 1.1, are illustrated below [5]:

- Step 1: All the AM processes start from a blueprint of the final end-user product which could be created using different methodologies,

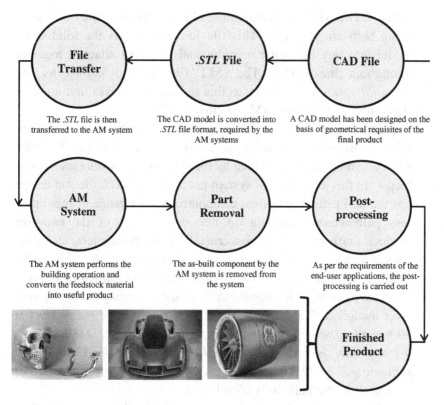

Figure 1.1 Procedural steps of AM.

primarily depending on the requirement. The use of a CAD software (such as CATIA®, Unigraphics™, SolidWorks®, ThinkerCAD, etc.) is one of the most widely used methodologies for the development of the engineering products. However, for biomedical applications, Computer Tomography (CT) and Magnetic Resonance Imaging (MRI) are the two most versatile approaches of collecting the patient specific data that can be further converted into a CAD file by using specialized medical software (such as MIMICS and 3D Doctor). In the case of reverse engineering applications, scanners are also used for producing the cloud database to obtain a CAD model.

- Step 2: Once the CAD model is mature, it is converted into *Standard Tessellation Language* (*.STL*) file format, preferred by the AM systems. The *.STL* file format can be obtained by using the CAD modelling software packages. This file format converts the solid CAD model into tiny triangular or polygonal networks attached together throughout the model. The *.STL* file is also known as the *Stereolithography* file format as this special format was first used for stereolithography-based AM systems in 1987.

- Step 3: The as-produced *.STL* file is then transferred to the AM system, either by using LAN or wireless network. The as-received file by the AM system is further processed by the supporting software interface.
- Step 4: In this step, the AM system processes the *.STL* file and develops the tool path. The system can control a wide range of input process parameters depending on the requirements of the end-user product. Furthermore, the most crucial process parameters, including layer thickness, printing speed, and in-fill density of the AM system can be controlled. The most important activity performed by the AM interface software is to slice the *.STL* model according to the required layer thickness. *For example, if the final thickness of the product is 10 mm and the build layer thickness selected by the AM system is 0.5 mm, then the total number of slices formed by the AM interface software will be 20.* Figure 1.2 presents a schematic representation of slicing along the build-axis (Z-axis).
- Step 5: Once the product has been completed by the system, it is then removed from the build zone. Any material stuck with the build model is removed by using light-handed tools.
- Step 6: The as-produced physical model is then post-processed to make it suitable for the end-user application. Most manufactured models undergo sanding, tumbling, high-pressure air cleaning, polishing, and colouring.

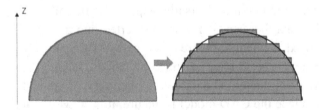

Figure 1.2   Slicing of the CAD model (left to right, adopted from [6]).

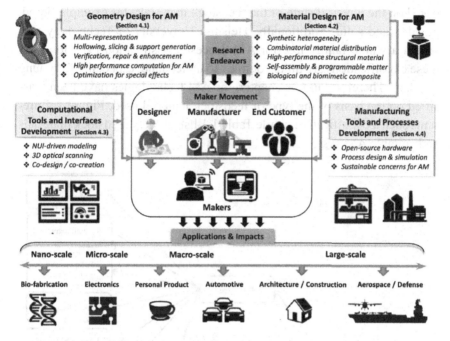

Figure 1.3 AM systems and subsystems (adopted from [7]).

- Step 7: This is the final stage where the manufactured component is ready for servicing.

Apart from the aforementioned seven-step series which apply to most of the AM systems, there could be some iterations in some of the application-specific systems. Further, it is possible that the dependencies of the AM systems on material modelling, design tools, computing, and process design can make it more challenging and cumbersome (refer to Figure 1.3).

1.4. AM Classification and Feedstock Materials

The classification of AM technologies has evolved enormously because of the extensive innovative breakthroughs undertaken since its inception in the late 20th century. J.P. Kruth, an eminent researcher of KU Leuven University, Belgium, classified the AM technologies in the 1990s [8] based on the physical state of the feedstock materials, for example liquid, powder, and solid. Additionally, Helsinki University of Technology presented a whole family tree for AM process classification [9]. However, these classifications are quite complex to understand, especially for young scholars. Therefore, in this chapter, the AM technologies have been classified on the basis of the type of feedstock material used. Figure 1.4 presents such a classification, while Table 1.1 lists the key feedstock materials and suitable AM systems.

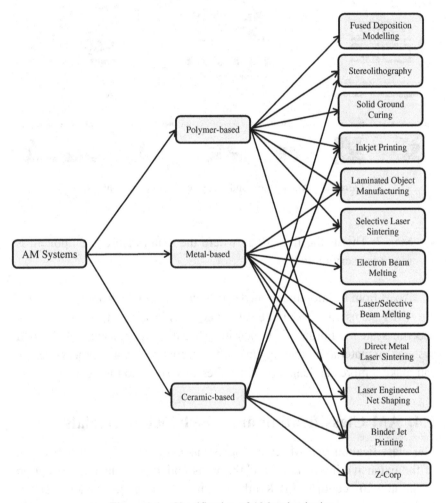

Figure 1.4   Classification of AM technologies.

Table 1.1   List of key feedstock materials and their applications.

Class of Materials	Material	Suitable AM Systems
Polymers	Thermoplastics filament wire (such as: acrylonitrile-butadiene-styrene, polylactic acid, polyamide, polyurethane, polycaprolactone, polyethylene, polyethylene glycol, poly-ether-ether-ketone, poly(methyl methacrylate), ULTEM, high impact polystyrene, and polyvinyl alcohol, etc.)	Fused Deposition Modelling
	Wax and acrylic liquid monomers	Inkjet Printing
	A wide range of ultraviolet curable liquid photopolymers	Stereolithography
		Solid Ground Curing
		Lithography
	Adhesively coated thermoplastics sheets	Laminated Object Manufacturing
	Thermoplastics powder (such as: polyamides, polystyrenes, thermoplastic elastomers, poly-aryl-ether-ketones, and polycarbonate)	Selective Laser Sintering
Metals	A wide range of metal/alloy powders (such as: aluminium, brass, bronze, cobalt chromium, copper, gold, platinum, steel, silver, titanium, Inconel, etc.)	
	A wide range of metal/alloy powders (such as: copper, niobium, aluminium, bulk metallic glass, stainless steel, titanium aluminide, cobalt chromium, Inconel, etc.)	Electron Beam Melting
	A wide range of metal/alloy powders (such as: stainless steel, maraging steel, cobalt chromium, Inconel, aluminium, titanium, etc.)	Laser/Selective Beam Melting
		Direct Metal Laser Sintering
		Laser Engineered Net Shaping
	Adhesively coated metallic sheets	Laminated Object Manufacturing
Ceramics	Powder of silicon carbides, aluminium oxides, boron carbide, zirconia, sand, diamond, etc.	Binder Jet Printing
	Plaster (ZP250)	Z-Corp®

Feedstock systems are primarily polymer-, metal-, and ceramic-based; however, there exist some specialized systems for manufacturing fibers, sand, plasters, glass, wood, and other biomaterials which are mainly used for research applications. It can be seen in Figure 1.4 that the maximum number of AM technologies exists for the polymer-based feedstock owing to cost effectiveness and ease of processing. Polymer-based AM technologies are also less expensive when compared with the metal- and ceramic-based technologies. In this particular class, fused deposition modelling is the only one that has been adopted for hobbyist activities. Metal-based AM technologies have been adopted for functional engineering applications, for example aerospace and automotive applications. Recent applications of metal AM technologies are biomedical implants and tools. Ceramic-based AM technologies have been least practiced for engineering applications, however these have been explored for developing metal casting moulds and customized scaffolds. Further, it can be seen from Table 1.1 that the variety of commercial printers is able to produce functional components by using a wide range of engineering materials. This is another reason behind the widespread popularity of the AM technologies in today's contemporary era of manufacturing.

1.5. Commercial Market of AM

The continuously developing legacy of AM resulted in the massive influx of competitive forces for developing the hardware, software, and feedstock materials for industrial manufacturing solutions of serial production, mass customization, and unique healthcare solutions. Indeed, the success of AM depends on how precisely and efficiently the AM products serve its intended market. The commercial success of AM also depends on how confidently the properties of materials can be translated into the desired shape or structure to meet predefined standards while maintaining competitive production costs. To boost the commercial market value of AM products, it is important to control and optimize the metrology by better utilization of expensive feedstock materials, catalysing production yield, eliminating part rejection, increasing energy efficiency, and decreasing post-processing requirements.

AM has been considered to be a market disrupter which means that it has the potential to replace current supply chains. Statistical reports indicate that within the first decade after the first patent had been granted in 1986, the market value of AM had grown to 1 billion USD. The rapid expansion of AM in its starting phase was because of the fact that the cost of the AM machine was 20 times higher as compared to today's scenario. The key reasons for the cost effectiveness of AM systems are the expiration of technology patents along with increasingly global outreach of AM systems from the US, Europe, and Canada. Some key milestones set by the AM market sector have been outlined [10].

With the passage of time, industrial leaders have begun to incorporate AM technologies as a cost-effective means of strengthening supply chains. In future, the continuous improvements and innovations in AM systems will be of paramount importance for their commercial uptake. However, it still needs to resolve the challenge of industrial scale implementation, and future directions are currently at a turning point. The industrial groups of North America stated that by 2020 they would use the Selective Laser Sintering systems for a large portion of their operations [11]. The survey report by Campbell *et al.* [12] indicated that Direct Metal Laser Sintering, Selective Laser Sintering, and Fused Deposition Modelling will be the top profitable AM systems by 2024 (refer to Figure 1.5). Furthermore, future growth of the AM market will also stand on the Factory 4.0 or Manufacturing 4.0 benefit from Digital Innovation

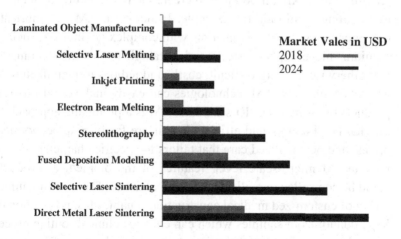

Figure 1.5 Market value of AM systems by 2024 (adopted from MaterksandMarkets. com).

Hubs and industrial ventures providing cost-effective AM systems for increasing their accessibility.

1.6. Applications of AM

Since the inception of AM technologies, its industrial applications have grown at a very high pace, merging into interdisciplinary branches of science and engineering. Nowadays, there exists a wide range of applications of AM technologies, as described in Figure 1.6. The development of AM in some of the prominent industrial sectors are because of the following reasons [12]:

- *Automotive manufacturers*: The reverse engineering assistance of AM helps the industry to get new products by saving time and development costs. The automotive companies are using AM for expanding their range of parts, including engines and vehicle bodies. It is increasingly being applied to lightweight metals and carbon fibers to make small quantities of structural and functional components (such as: engine exhausts, drive shafts, gear box components and braking systems for luxury).
- *Aerospace companies*: AM technologies fabricated highly complex and high performance aerospace products by eliminating assembly features and making it possible to create internal functional channels and cooling manifolds for example. In addition, AM can fabricate components of advanced materials with complex geometries, such as titanium, nickel, special steels and ultrahigh temperature ceramics, that otherwise are very difficult, costly and tedious to manufacture.
- *Medical industries*: AM technologies can easily make solid medical products by using CT/MRI scan data. AM is a promising approach to provide good quality and efficient healthcare facilities at economical prices, and personalized care that tailors to specific characteristics of patients. At micro-scale level, features of the biomedical products could be controlled to replicate the original tissue. The rapid manufacturing of customized medical constructions enables fast production of large quantities of samples which can enhance clinical routine procedures in terms of better meeting daily surgical needs. The recent

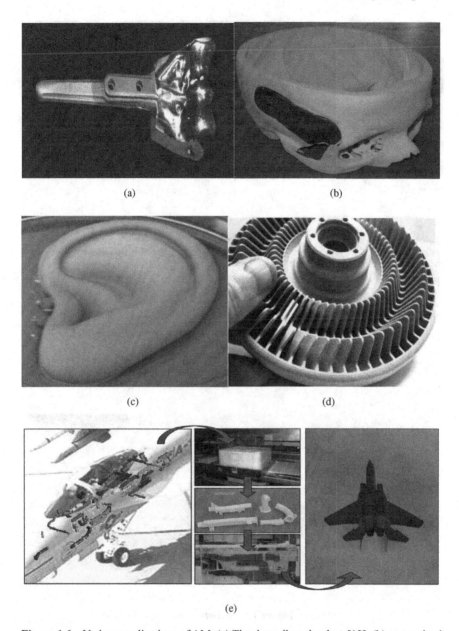

Figure 1.6 Various applications of AM. (a) Titanium elbow implant [12], (b) customized bone substitute [13], (c) artificial organ [14], (d) aerospace turbine model [12], (e) air-cooling ducts of aircraft [15], (f) AM Urbee car [16], (h) AM house [17], (i) jewellery item [18], (j) textile dress [19], (k) printed food [20], and (l) AM of tablets [21].

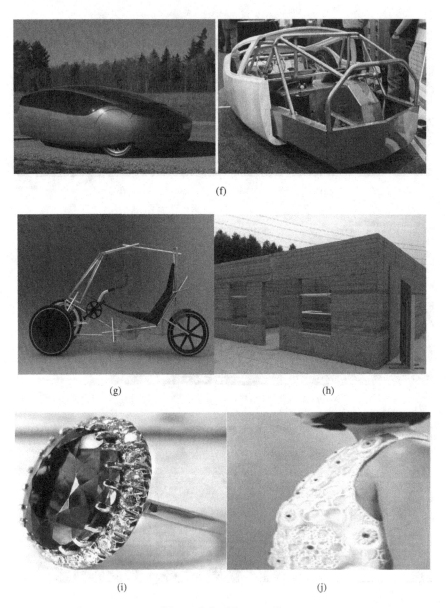

Figure 1.6 (*Continued*)

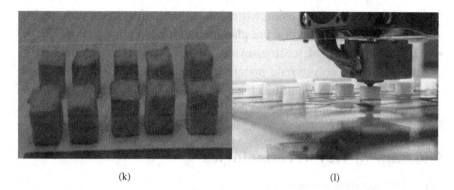

(k)   (l)

Figure 1.6 (*Continued*)

applications of AM suggest that artificial liver, kidney, heart, and lung could also be printed using human DNA cells. In pharmaceutical sectors, AM has been used for the synthesis of formulated medicines.

- *Machine tool industry:* AM facilitates on-demand manufacturing opportunities by reducing raw material usage and energy consumption. The AM processes can be used to attain true 3D microproducts. There is no need of tooling for production of spare parts, so it is unnecessary to hold legacy tooling in storage.
- *Construction and architectural industry:* The AM technologies can strengthen the abilities of the architects to create physical models faster without worrying about the complexity of their design. Furthermore, it helps in achieving better resolution than other conventional processes used in architecture. This digital fabrication means it is producing buildings with freeform surfaces that cannot be built by any known method today.
- *Jewellery:* Using CAD software, AM techniques can fabricate a variety of jewellery items as it is extremely capable of working with precious metal-based core components with required strength as compared to welding. It is now also possible to fabricate designed architectures of jewellery items which are otherwise not feasible by even a skilled jeweller.
- *Printed Food:* AM nowadays is producing blended edible items (such as pasta, sausages, breadsticks, ice creams, chocolates, and certain

breakfast cereals) by using extrusion technologies. It not only maintains the taste of the food but also their consistency.

* *Textiles:* With AM, custom-made clothing could be created using thermoplastic polystyrene material. Commercial modelling software can create clothing that suits the custom fit of a human body by tailoring the frames of rigid or flexible polymers.

Glossary of Key Terminologies

3D: Three-dimensional
AM: Additive Manufacturing
CAD: Computer-Aided Design
CT: Computer Tomography
DNA: Deoxyribonucleic Acid
MRI: Magnetic Resonance Imaging
RM: Rapid Manufacturing
RP: Rapid Prototyping
STL: Standard Tessellation Language
USD: United States Dollar

Exercise Questions

Q1: What is additive manufacturing?
Q2: Differentiate between additive and subtractive manufacturing.
Q3: Describe the principle and working procedure of additive manufacturing in detail.
Q4: Give the classification of additive manufacturing technologies on the basis of type of feedstock material.
Q5: What are the various applications of additive manufacturing?
Q6: Write a short note on the history of additive manufacturing.
Q7: Discuss the market of additive manufacturing in brief.

References

1. Wohlers T, Gornet T. History of additive manufacturing. *Wohlers Report.* 2014;24(2014):118.

2. Pham DT, Dimov SS. Applications of Rapid Prototyping Technology. In Rapid Manufacturing 2001 (pp. 87–110). Springer, London.

3. ASTM. 2015. Standard terminology for additive manufacturing — general principles. Part 1: Terminology. ISO/ASTM Stand. 52792, ASTM

4. Gibson I, Rosen DW, Stucker B. Additive Manufacturing Technologies. New York: Springer; 2014.

5. Huang SH, Liu P, Mokasdar A, Hou L. Additive manufacturing and its societal impact: A literature review. *The International Journal of Advanced Manufacturing Technology*. 2013 Jul 1;67(5–8):1191–1203.

6. Zha W, Anand S. Geometric approaches to input file modification for part quality improvement in additive manufacturing. *Journal of Manufacturing Processes*. 2015 Oct 1;20:465–477.

7. Gao W, Zhang Y, Ramanujan D, Ramani K, Chen Y, Williams CB, Wang CC, Shin YC, Zhang S, Zavattieri PD. The status, challenges, and future of additive manufacturing in engineering. *Computer-Aided Design*. 2015 Dec 1;69:65–89.

8. Kruth JP. Material incress manufacturing by rapid prototyping techniques. *CIRP Annals*. 1991 Jan 1;40(2):603–614.

9. Koukka H. The RP family tree, Helsinki University of Technology, Lahti Centre. URL: http://shatura.laser.ru/rapid/rptree/rptree.html.

10. Tofail SA, Koumoulos EP, Bandyopadhyay A, Bose S, O'Donoghue L, Charitidis C. Additive manufacturing: Scientific and technological challenges, market uptake and opportunities. *Materials today*. 2018 Jan 1;21(1):22–37.

11. The 3-D Printing Revolution, https://hbr.org/2015/05/the-3-d-printingrevolution.

12. Campbell RI, De Beer DJ, Pei E. Additive manufacturing in South Africa: Building on the foundations. *Rapid Prototyping Journal*. 2011 Mar 8;17:156–162.

13. Ciocca L, De Crescenzio F, Fantini M, Scotti R. Rehabilitation of the nose using CAD/CAM and rapid prototyping technology after ablative surgery of squamous cell carcinoma: a pilot clinical report. *International Journal of Oral & Maxillofacial Implants*. 2010 Aug 1;25(4).

14. https://interestingengineering.com/the-science-fiction-world-of-3d-printed-organs

15. Lyons B. Additive manufacturing in Aerospace; examples and research outlook, frontiers of engineering. URL: http://www.naefrontiers.org/File.aspx?id=31590, 2011.

16. http://green.autoblog.com/photos/sema-2010-kor-ecotec-urbee-3536115/.

17. http://www.winsun3d.com/En/About/

18. https://www.spilasers.com/application-additive-manufacturing/additive-manufacturing-in-the-jewellery-industry/

19. https://www.sculpteo.com/en/applications/textile-industry/

20. Lipton JI, Cutler M, Nigl F, Cohen D, Lipson H. Additive manufacturing for the food industry. *Trends in Food Science & Technology*. 2015 May 1;43(1):114–123.
21. https://3dprintingindustry.com/news/fabrx-launches-the-m3dimaker-for-personalized-medicine-3d-printing-170695/

Chapter 2

Polymer-Based Additive Manufacturing System

This chapter introduces the different types of polymer-based Additive Manufacturing (AM) technologies by emphasizing on their work principles, feedstock materials, and potential applications.

2.1. Polymer

The term "polymer" is derived from the Greek words *polus*, meaning "many", and *meros*, meaning "part". Therefore, the combined term refers to a molecule which is made of repeated similar parts known as "monomers". The monomers are converted into a giant polymer chain through a chemical process, known as polymerization. There are mainly two different types of polymerizations: addition polymerization and condensation polymerization.

A. Addition Polymerization
In this case, a chemical chain reaction is started by providing a small amount of initial energy that converts monomers into polymers. It includes three steps, namely initiation, propagation, and termination. The addition of the energy initiates the reaction that further propagates until the required volume of the polymer is obtained. Finally, the

reaction is terminated. To understand it better, let us take an example of polyethylene. For this, the reagents are ethylene monomer and free radicals, as given below:

- Ethylene Monomer

- Peroxo compounds are a good source of free radicals, which are chemical species containing unpaired valence electrons that can form a covalent bond with an electron on another molecule. Peroxo compounds are inherently unstable and readily undergo homolytic cleavage to form alkoxy radicals.

*Indicates free electron

i. Initiation: This involves the reaction between the ethylene monomers and free radicals.

ii. Propagation: This involves the addition of another monomer.

iii. Termination: In this stage, another free radical has been provided to the chain to end further propagation.

B. Condensation Polymerization

In the case of condensation polymerization, polymers with a molecular weight lower than the sum total of their individual molecular weights have been produced. It involves two different types of end groups reacting with each other and forming a chain along with the loss of a small molecular by-product. To understand it better, let us assume an example of polyamide.

adipic acid   hexamethylene diamine

nylon 6,6

2.1.1. *Types of Polymers*

Polymers can be classified on the basis of their behaviours, for example under heating. Followings are the three major types of polymers:

A. Thermoplastics

These are the polymers which have strong covalently bonded chains and weaker secondary *van der Waals* bonds. This enables them to get recycled by thermal decomposition. A few common applications of thermoplastics are bottles, cable insulators, tape, medical syringes, mugs, textiles, packaging, and insulation.

B. Thermosets

Thermosets have stronger cross-linkages that allows the material to resist softening upon heating. Due to its excellent thermal and flame retardancy, it is used for making fry pan handles, electric circuits, and other parts where insulation is required.

C. Resins/Elastomers/Waxes

In polymer chemistry, resin is a highly viscous natural or synthetic substance that can be converted into polymers. Generally, resins are mixtures of multiple compounds. On the other hand, elastomers are viscoelastic polymers with very weak intermolecular forces. Waxes are organic compounds that are malleable near ambient temperatures.

2.1.2. *Some Key Terminologies*

- *Molecular weight:* It is the weight of a given molecule in Daltons. Due to the isotropic nature of a material, its different molecules may possess different molecular weights.
- *Cross-linkage:* It is the bonding of a one polymer chain with another.
- *Polymer blend:* It is the mixing of more than one polymer.
- *Polymer additive:* These are the chemical compounds that are used in polymers for improving their properties.
- *Feedstock:* These are the raw materials, in the required physical state, for producing end user products.
- *Glass transition temperature:* It is the temperature at which polymers start to behave like elastic materials because of the weakening bonds by heating.
- *Curing:* It is a chemical process to obtain required toughness and strength.

2.2. Polymer Feedstock

Polymers are considered as the most common materials in the AM industry owing to their diversity and ease of adoption to different 3D printing processes. The polymer feedstock materials of AM systems, majorly driven from petroleum raw materials and natural resources, possess low production cost, excellent flexibility, and provide an efficient mode of replacing metals and alloys for producing high-quality products for a myriad of engineering and scientific applications. Polymeric materials possess significantly higher elasticity as compared to conventional metals and ceramics. This is because the atomic structure of the polymers enables ease of dislocation upon loading. Referring to Figure 2.1, it can be seen that the atomic structure of a polymer consists of both crystalline (straight lines) and amorphous phases (irregular line) connected together with weak atomic forces provided by ionic and covalent bonding. However, in the

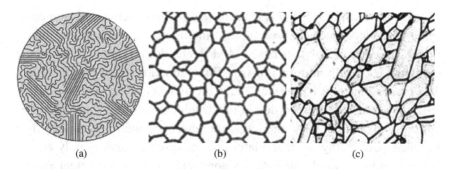

Figure 2.1 Atomic structures of polymer (a), metal (b), and ceramic (c).

case of metals, the atoms, arranged in a regular pattern, share delocalized electrons to obtain stronger metallic bonding. Owing to this, the mechanical strength of metals is higher than polymers, while the elasticity is compromised. Lastly, in the case of ceramics, the atoms are held together by ionic and covalent bonding that is much stronger than in metals. Therefore, they resist, enormously, to the externally applied load which often results in sudden failure without showing elastic deformation.

Polymer-based materials have the capability to soften and attain desirable rheological flow upon heating, while regaining its firmness upon cooling. In addition to this, polymer feedstock can be easily mixed with other polymers or fillers or reinforcements to create blends or composites which are utilized in the manufacture of high-performance engineering thermoplastics. There exists a wide variety of polymer materials which have substantial mechanical properties (such as tensile and compression strength and heat resistance), biocompatibility, biodegradability, geometrical integrity, water solubility, ease of colouring, etc. Today, polymer-based feedstock systems are rapidly evolving and breaking the limits of AM technologies. Polymers for AM are found in the form of thermoplastic filaments, reactive monomers, resin or powder. The capability of employing 3D printing of polymers has been explored for several years in many industrial applications, such as the aerospace, architectural, toy fabrication and medical fields. Driven in the long term by massive opportunities in AM production, the polymer feedstock industry is believed to generate $11.7 billion of revenue in 2020 including sales-developed polymers and associated hardware [1].

AM's value proposition is clear for automotive, aerospace, and consumer products (such as prototypes, moulds, casts, tooling, and functional products). Polymer-based AM technologies are continuously evolving, especially in terms of workflow automation and optimization. The most significant progresses and achievements of the polymer-based AM market have been recorded in the development of new materials for currently available technologies. Emerging polymeric materials, for instance poly-methyl-methacrylate, poly-ether-ether-ketone, poly-ether-ketone-ketone and poly-aryl-ether-ketone, are the most competent examples of high performance thermoplastic materials for existing AM systems that have been widely acknowledged for their high heat resistance, chemical resistance, biocompatibility, and ability to withstand high mechanical loads.

Following the unique "Additive" principle, the systems build the polymeric objects by adding the polymer layer-upon-layer. Different types of polymer-based systems adopt different procedures for building an object, as given below:

- *Melt fusion or Extrusion:* In this procedure, the AM system melts and fuses the polymer feedstock, normally available in the form of filament wire, by using in-built heating filaments. The intensity of the heat supplied here matches, perfectly, with the required rheological flow of the fused material from the die or passage. The solidification of the deposits takes place automatically, with time.
- *Jetting:* Being distinct from melt fusion, this procedure adopts the jetting procedure to deposit the polymeric solutions, from the tiny nozzles, onto the platform, The curing of the deposits take place by using ultraviolet (UV) lights, an integrated part of the AM system. It adopts a photopolymerization-based chemical process.
- *Binding jetting:* This procedure is quite similar to the jetting procedure as the binding agent has been deposited from the tiny jet-ways onto the solid powder particles to impose the green strength. However, post-curing is necessary to impart mechanical strength to the produced object.
- *Lithography:* This procedure also adopts a photopolymerization process; however, the required polymer solution is not injected on the

platform, like in the previous case, but is already contained in a reservoir. The high-powered UV laser strikes onto the liquid resin solution and cures to harden it.

- *Sintering:* In sintering, the powder-based polymeric feedstock system is heated and welded together, by using a high intensity laser beam, to form a green product. This approach is also referred to as an indirect AM process as the green product often requires further curing to obtain desirable mechanical and structural integrity.
- *Stacking:* Unlike the previous procedures, the stacking method uses an adhesive-coated polymer sheet which can be activated by heating. Therefore, the different stacks are joined with one another through heating, while the final shape is given by a laser cutter, also known as knife.

Since there exists different work procedures, the type, form, and class of polymeric feedstock materials vary extensively; refer to Table 2.1 for better understanding.

2.3. Polymer-Based AM Systems

The implications of polymer-based AM systems, for both industrial and general public use, have grown enormously in recent years. This rapid evolution of the market has placed polymer-based AM systems not only in tremendously varied industrial settings, but also in schools, public libraries, universities, laboratories, and more commonly now, in homes [2]. These systems, in context of the whole AM class, offer the following key benefits:

- Processing of goods through these systems is cost-effective as compared to metal and ceramic AM systems.
- The capital cost of these systems is also less as compared to metal and ceramic AM systems.
- This class of AM technologies enjoys the benefit of a maximum number of commercial systems suiting a wide range of industrial applications.

Table 2.1   List of commercially available polymer feedstock for AM systems.

		Procedure						
		Melt-fusion	Jetting	Binder-jetting	Lithography		Sintering	Stacking
State	Material	FDM	IP	BJP	SLA	SGC	SLS	LOM
Liquid	Epoxy resin		✓	✓	✓	✓		
	Acrylic resin		✓	✓	✓	✓		
	Binder/powder hybrids		✓					
Powder	Polyamide						✓	
	Polycarbonate	✓					✓	
	Polystyrene	✓					✓	
	Acrylonitrile-butadiene-styrene	✓						
	Polypropylene						✓	
	Starch		✓	✓				
	Elastomer/Cellulose		✓	✓				
	Polylactic acid	✓	✓	✓				
	Thermoplastic polyurethane	✓						
	Polyethylene	✓						
	Poly-ether-ether-ketone	✓						
Solid sheet	Polyester film							✓
	Polyolefin film							✓
	Polyvinyl copolymer film							✓
	Thermoset film							✓
Molten liquid	Acrylonitrile-butadiene-styrene	✓						
Filament	Acrylonitrile-butadiene-styrene/ polycarbonate blend	✓						
	Polyphenylene sulfine	✓						
	Polyamide	✓						
	Polycarbonate	✓						
	Polystyrene	✓						
	Acrylonitrile-butadiene-styrene	✓						
	Polypropylene	✓						
	Polylactic acid	✓						
	Poly-ether-ether-ketone	✓						

Note: FDM, IP, BJP, SLA, SGC, SLS, and LOM refers to fused deposition modelling, inkjet printing, binder-jet printing, stereolithography, solid ground curing, selective laser sintering, and laminated object manufacturing, respectively.

- The manufacturing flexibility of polymer AM systems is extremely high.
- Customization of feedstock materials is also feasible for experimentation and exploration of new applications.

Following are the technical insights of polymer AM technologies, along with their key benefits, limitations, and potential applications.

2.3.1. *Fused Deposition Modelling*

Fused deposition modelling (FDM), also known as fused filament fabrication (FFF) and extrusion fabrication, is perhaps the most common AM technology available nowadays. The main credit to this goes to the low cost and robustness of this system. This AM method utilizes filament wires of thermoplastic materials to fabricate parts. However, there exists certain modified FDM systems that employ feedstock available in pellet or granular form. For manufacturing, the FDM system raises the temperature of the feedstock material above glass transition temperature, a temperature at which the plastic behaviour of the material changes to viscoelastic behaviour, to enable the flow of the melt. Finally, the heated material is extruded through a nozzle onto a build platform [3]. Figure 2.2 shows the schematic representation and actual pictorial description of the FDM system. Further, a much appreciated application of the FDM system, especially by the biomedical industry, is the pressure-assisted micro-syringe system (refer to Figure 2.2(c)). This system has been extensively used for tissue engineering to create soft tissue scaffolds as well as in pharmaceutical industries for producing gel- or paste-based medicines.

As can be seen from the schematic representation, the filament wire of a thermoplastic material is drawn from the spool by the feed rollers and passedt to the print-head. The print-head is the place where the thermoplastic material is heated at the desirable temperature, which mainly depends on the thermal properties of the material, and then pushed further to escape through the tiny nozzle at the bottom of the print-head. The size of the extruded slice depends on the size of the nozzle used, and it affects the surface and mechanical characteristics of the resulting component. The extruded slice is then placed on the fixtureless platform, where it solidifies.

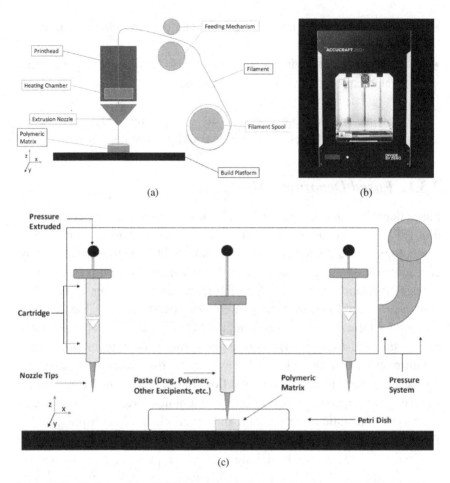

(a)

(b)

(c)

Figure 2.2   (a) Schematic of FDM system [3], (b) pictorial view of the FDM machine (adopted from DividebyZero.com), and (c) pressure-assisted micro-syringe system [3].

The process continues by following the "layer-upon-layer" approach until the last slice has been placed at the system defined location.

It is important to understand that there are three different motions along the X, Y, and Z axes. The motions of the FDM system components, mainly print-head and build platform, have been automatically generated by the software interface provided by the vendor along with the hardware. Depending on the geometry of the computer-aided design (CAD) of the solid model, the software interface develops a tool path obeyed by the print-head and build platform. Normally, the print-head can move along the X and Y axes, while the build platform can move along the Z axis. When a slice is placed on the build platform, it lowers down by a height equivalent to the selected layer thickness that further depends on the size of the nozzle used. In this way, any critical three-dimensional (3D) shape of the CAD model can be converted into the physical component. However, it is an intrinsic feature of the FDM system that the as-manufactured physical component possesses *internal porosity* and *external staircasing*. The internal porosity can be defined as the production of void spaces between the adjoining slices of the fused thermoplastic, while the external staircasing is the formation of steps on the exterior of the fabricated object. Figure 2.3 illustrates the internal porosity and external staircasing of the FDM build products.

Compared to other AM methods, FDM is regarded as a *"clean"* method as it uses no powder or liquid polymer. However, for the over-handing structures (such as T, O, and inverted U), the system utilizes the support material provided by support material print-head in a way similar to the model material print-head. The system enables syncing between these two print-heads so that depending on the requirement, the desirable material could be printed. Furthermore, the composition of the support

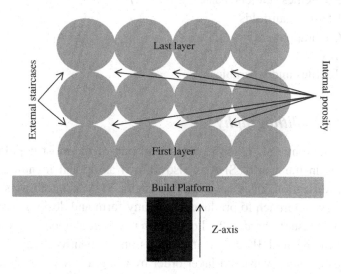

Figure 2.3. Representation of internal porosity and external staircasing.

material is formulated such that it can dissolve in water or higher pH solutions. Many commercial manufacturers these days are developing methods of direct printing for glass and carbon fibers. Followings are the key benefits, limitations, and potential applications of the FDM system:

- Benefits of FDM
  - ○ Reasonable cost
  - ○ Less operational time
  - ○ Broad range of feedstock availability
  - ○ Ease of use
- Limitations of FDM
  - ○ Geometrical anisotropy of the products
  - ○ Requires support structures
  - ○ Limited build size
  - ○ Warpage and thermal deformations of products
- Applications of FDM
  - ○ Prototyping
  - ○ Machine components
  - ○ Inserts and tools
  - ○ Orthopaedic products (scaffolds, tissues, etc.)
  - ○ Medicines (tablets, pastes, gels, etc.)
  - ○ Engine manifolds
  - ○ Casting patterns
  - ○ Toys
  - ○ Textiles and wearables

2.3.2. Stereolithography

Stereolithography (SLA) is the oldest and one of the most popular AM techniques in the world. SLA is also known as optical manufacturing, photo-solidification, or resin printing system. SLA technology is a fast and effective approach to produce nearly any form and design. Basically, it uses a UV laser to harden the liquid resin that is contained in a reservoir to create the desired 3D shape. It converts photosensitive liquid into 3D solid plastics in a layer-upon-layer order by using a low-power laser and photochemical processes that allow light to link chemical monomers and oligomers to create polymers. The SLA method has been primarily utilized to produce templates, prototypes, designs, and manufacturing elements. Every standard SLA 3D printer is generally composed of four primary sections:

- The tank is filled with the liquid resin
- A perforated platform is immersed in a tank
- High-powered UV laser strikes on the resin
- The computer interface manages the platform and the UV laser movements

Figure 2.4 represents the SLA process. As can be seen, the UV laser source can move along the X and Y axes, while the liquid resin chamber moves along the Z axis. As the processing starts, the UV laser irradiates the resin and creates the two-dimensional (2D) layers in subsequent steps. As the first layer of the resin cures, the liquid resin chamber lifts up and fresh resin comes on the top of the previously cured resin to process the second layer. This continues until the desired height of the CAD model has been processed. Similar to FDM, the tool path to the UV laser source and the liquid resin chamber has been provided by a software interface.

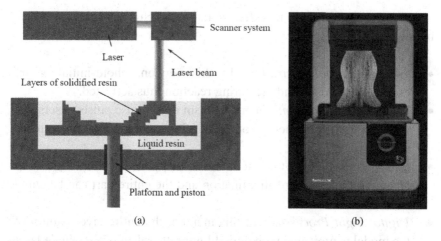

(a)   (b)

Figure 2.4 (a) Schematic [4] and (b) pictorial view (adopted from https://formlabs.com/asia/) of the SLA process.

The height of the cured layer has been defined by the software interface and accordingly the stepwise lifting of the liquid resin chamber takes place. In SLA, the cure depth is determined by the energy of the light to which the resin is exposed, which can be further controlled by adjusting the power of the light source, the scanning speed or the exposure time [5]. The as-produced polymeric component is taken out of the build chamber and washed away to remove the uncured traces of the liquid resin. Indeed, the SLA system is more expensive than FDM and is considered as less user-friendly. Owing to this, SLA is less common for hobbyists and for 'desktop' printing. Furthermore, it is also a slower process than FDM and is impractical for large objects. However, SLA has comparatively higher resolution and the fabricated parts possess a layer thickness of ~20 μm. Moreover, the surface roughness and final appearance of SLA parts are generally better than the FDM parts along with greater geometrical accuracy. Following are the essential requirements of resins used in SLA systems:

- The resin must have finetuned viscosity to facilitate smooth and stable surfaces. This requirement has limited SLA resins to acrylates.
- The resin should be highly reactive to print the structures in a reasonable amount of time.

There exists various modified versions of SLA systems, as given below [6]:

- *Two-photon polymerization:* In this version, photo-initiators are excited in the resin and the curing reaction thus activated.
- *Pinpoint solidification:* The mechanism doesn't use pulsed lasers and with focused polymerization resolutions under 0.4 μm can be achieved.
- *Bulk lithography:* In this method, 3D components are produced by changing the energy of illumination and the entire part can be cured all at once.
- *Digital Light Processing:* In this method, the entire layer pattern of the model is projected with digital light rather than with a single laser point. This increases the manufacturing speed and makes it more affordable.

- *Free surface approach:* In this version, the building platform of the part, located in the resin tank, is covered by liquid resin film. After curing of the first layer in the resin tank, a new resin layer is supplied by a mechanical sweeper and, simultaneously, the base with the resin tank is lowered along the Z-axis.
- *Constrained surface approach:* This bottom-up approach is entirely different. The UV laser, in this case, strikes on the surface from below the resin tank and after curing of each layer the tank lifts up rather than down, as in the case of normal SLA.

The benefits, limitations, and applications of SLA are given below:

- Benefits of SLA
  - SLA is one of the most precise techniques
  - Products created are of high quality and finetuned
  - Tightest dimensional tolerances
  - Smooth surface
- Limitations of SLA
  - Manufacturing times are long
  - Resins are fragile
  - Resins are proprietary and un-exchangeable
  - Printing cost is high
- Applications of SLA
  - Prototyping
  - Machine components
  - Orthopaedic products (scaffolds, tissues, etc.)
  - Medicines (tablets, pastes, gels, etc.)
  - Toys
  - Casting patterns
  - Textiles and wearables

2.3.3. *Solid Ground Curing*

Solid ground curing (SGC), as given in Figure 2.5, is one of the oldest AM systems. While the strategy offered great exactness and a high creation rate, it experienced high securing and working expenses because of framework unpredictability. This prompted poor popularity of this system when

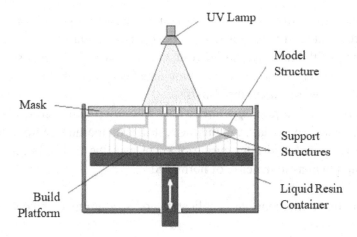

Figure 2.5   Schematic representation of the SGC process.

compared to other competent polymer AM systems. The SGC technique is a photography-based approach that adds the subsequent layers of polymer resin by using a UV lamp, rather than laser as in the case of SLA. The step-by-step working procedure of the SGC process is discussed below:

- In the first step, the CAD model is drawn on a glass plate. In this way, the surface to be solidified is kept transparent, while the rest of the glass sheet is covered with colour by using an electrostatic process.
- The glass plate is then placed between the liquid resin and a UV lamp. The uncoloured regions of the glass plate allow passage of the UV light onto the resin, causing it to solidy.
- The non-illuminated surface remains as it is.
- After this, the glass plate is removed to be coloured according to the next layer, and again placed between the liquid resin and UV lamp.
- The non-solidified monomer is wiped off.
- The in-filled unprocessed resin provides the support structures required.
- Once the parts are complete, the resin structures can be recovered.

Following are the benefits, limitations, and applications of the SGC process:

- Benefits of SGC
   ○ Does not need additional support material

- o   High accuracy along Z axis
- o   The printed masks are of one-time use
- o   No post-cure required
- Limitations of SGC
  - o   High operational cost
  - o   Creates waste
  - o   Milling is also sometimes required to finish the cured layers
  - o   SGC is a loud manufacturing process
- Applications of SGC
  - o   Prototyping
  - o   Machine components
  - o   Casting patterns
  - o   Toys

2.3.4. *Inkjet Printing*

Inkjet printing (IJP) technology is regarded as one of the fastest, most flexible, versatile, and cost effective AM processes that can fabricate both simple and complicated 3D objects without using any type of additional tooling. The basic idea of inkjet printing is derived from desktop printers where the feedstock ink is deposited on the paper as per the requirement. Similar to this concept, the IJP AM system deposits the formulated polymer-based inks onto the fixtureless platform which can move along the Z axis to provide an additional dimension to the construct [7]. The inkjet printer, as shown in Figure 2.6, begins with the build thermoplastic solution-based ink (model material) and wax (support material). These materials are fed to an inkjet print head which moves in the X and Y axes and shoots tiny droplets to the required locations to form one layer of the part. The model and support materials cool, instantly, and then a milling head moves across the layer to smooth the surface. The debris produced is vacuumed away by the particle collector. The elevator and build platform are then lowered down along the Z axis and the next layer building takes place in the same way. Once the process is completed, the part is removed and the wax support material is melted away.

In inkjet printing, the 3D components are produced with the highest dimensional accuracy and smoothness of the surfaces. Inkjet printing also allows different thermoplastic and biomaterials to be printed in the same

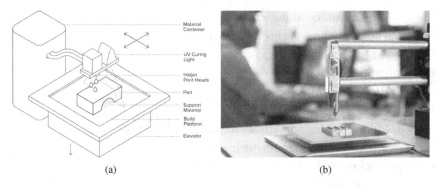

(a)   (b)

Figure 2.6   (a) Schematic representation of IJP (adopted from 3D Hub) and (b) pictorial view of IJP (adopted from Nano Di).

object, making it one of the most exceptional and practical AM technologies. Compared to the other AM technologies, lightweight supports are only required in areas with severe overhangs. However, the overall amount of material is almost identical to the entire build volume of the part, which makes the method less economical. Nevertheless, this technology enables the building of 3D multi-material multi-colour structures that otherwise are difficult with SLA, SGC, and FDM methods [8].

Following are the benefits, limitations, and applications of the IJP process:

- Benefits of IJP
  - Low cost as printing heads are readily available and easily manufactured
  - Fast build speeds
  - Multi-colour prints
  - High resolution and smooth surface finish
  - Large area printing
  - Part isotropy
- Limitations of IJP
  - Limited material choice of photopolymers and waxes
  - Low strength
  - Support structures required

- Applications of IJP
  - ○ Electronic components, OLEDs
  - ○ Biological structures
  - ○ Sensors
  - ○ Formulated medicines

2.3.5. *Laminated Object Manufacturing*

Laminated object manufacturing (LOM) was developed by Helisys of Torrance, CA, more than 25 years ago. However, the technology has also seen market depletion and is not well acknowledged as compared to FDM, SLA, and inkjet printing technologies. Owing to this the market share of LOM is very limited, as described in Chapter 1. Figure 2.7 shows the schematic description of the technology. Unlike the other processes, the technology uses laminates or sheets of adhesive-coated materials, which can also be of metal, papers, and ceramics. This highlights that LOM is highly flexible in terms of feedstock materials. Although LOM is

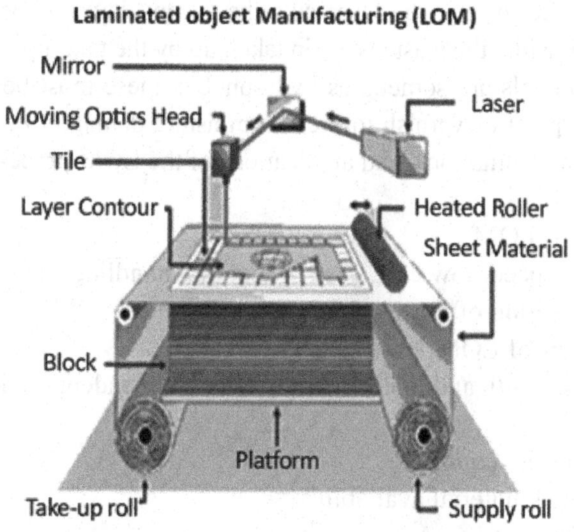

Laminated object Manufacturing (LOM)

Figure 2.7 Schematic representation of LOM (adopted from Metal-AM).

attractive, its downsides are the requirement of intensive labour, long processing times, and potential damage of the green parts. LOM manufacturing uses a similar layer-upon-layer approach but uses feedstock as material and adhesive as compared to welding. Further, the process utilizes the cross-hatching method during printing in order to allow easy removal after building. The prominent applications of this technology lie in producing objects for aesthetic and visual appearances and are not suitable for structural components.

Referring to Figure 2.7, the original material laminate sheet is supplied from one side by the supply roller and held by the take-up roller on the other side. The laser source can strike on the laminated sheet by using moving optics head (having X and Y axes movement) to cut the laminate sheet as per the requirements of the design of the CAD model. The cut sheet is placed onto the platform using the cross-hatch, while the waste sheet is taken up by the take-up roller. Cross-hatching breaks up the extra material, making it easier to remove during post-processing. This is the first layer printing step. The printing bed comes downwards (along Z axis) to the fixed height equivalent to the laminate sheet thickness. Afterwards, the fresh sheet comes on top of the first layer and the heated roller moves on the entire section to activate the glue. Again, the laser cuts the second layer as per the CAD model design, while the waste is again taken up by the take-up roller. As the feedstock materials are sometimes hydrophobic, these must be sealed and finished with paint or varnish to prevent moisture damage [9]. Followings are the benefits, limitations, and applications of the LOM process:

- Benefits of LOM
  - Good speed, low cost, ease of material handling
  - Fast cutting of the sections
- Limitations of LOM
  - The strength and integrity of models is dependent on the adhesive used
  - Labour is required
  - Limited material availability
- Applications of LOM
  - Prototypes
  - Artistic models

2.3.6. *Selective Laser Sintering*

Selective laser sintering (SLS) has been regarded as one of the most efficient and functional application-specific AM technology that has been extensively used for the production of polymer, metal, and composite products with substantial mechanical strength, geometrical accuracy, and attractive aesthetic appearance. SLS utilizes a high energy laser source (CO_2, Nd:YAG, Argon, etc.) to sinter the powdered plastic material into a solid structure. The low cost of production, high productivity, and established materials for this technology make it an ideal choice for a wide range of engineering applications, including prototyping.

Figure 2.8 shows the schematic of the SLS process. The process starts with CAD model-based instructions given to the system that include the tool path for both laser source and build part. As there are two movable platforms, the left side is used for providing the raw powder feedstock while the right side holds the powder as the high intensity laser strikes. The left- and right-sided platforms move up and down, respectively, during the processing. After the tool path is generated, the supply roller slides the powder particles from the left to the right-sided platform. Then the high power laser provides the thermal energy required for the powder

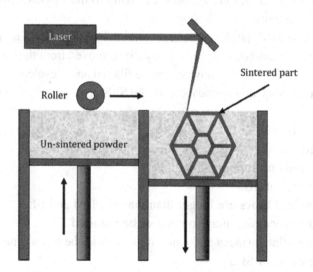

Figure 2.8   Schematic representation of SLS [2].

sintering. A beam deflection system is generally used here to focus the laser beam to the desired position, while scanning each layer.

After the first layer of the powder polymer particles sinters, the left-sided platform lifts up and the right-sided platform lowers down. Noticeably, the height by which the platforms move is equal to the layer height of the manufacturing object. Again, the new layer of powder particles is sintered onto the first layer using the same laser power. This process repeats until the desired 3D part is achieved. Technically, the sintering between powder particles occurs because their temperature is raised above the melting point of feedstock powder. Sometimes, when it is extremely difficult to attain the bonding of powder particles, additional binder materials are used.

The quality of the manufactured object depends on a wide range of processing parameters such as powder composition and morphology, laser energy input, scan speed and processing temperature. It is very important to calculate the laser power required for melting as excessive energy provided can result in distortions of the build product. Furthermore, scan speed and duration for which the laser strikes on the particles are also critical [10]. This process has great benefit in terms of the support structures which have been ingeniously built such that the loose powder particles after the manufacture can easily be collected and reused. It is recommended to use spherical particles owing to their ease of flow, dense packing, and required distribution.

The SLS-based product requires minimal post-processing time and labour. The as-manufactured product could be removed from the build chamber and cleaned. Any excess powder can be filtered and recycled. The various benefits, limitations, and applications of the SLS process are given below:

- Benefits of SLS
  - Parts have good isotropic mechanical properties
  - It needs no support structures
- Limitations of SLS
  - The lead times are longer than that of FDM and SLA
  - Post-processing may sometimes be required
  - Large flat surfaces and small holes cannot be printed accurately
- Applications of SLS
  - Prototypes
  - Tooling, inserts, and fixtures

- o Aerospace and military hardware
- o Electronic packaging devices
- o Orthopaedic tools and devices
- o Tissue engineering
- o Casting patterns

2.3.7. *Binder-jet Printing*

Binder-jet printing (BJP) is another excellent invention of the AM technology that can fabricate polymer, metallic, and ceramic components with a huge diversity of engineering materials. This technology is a combination of IJP and SLS, as the binder material is supplied following the IJP procedure and the model material is supplied similar to SLS technology. The BJP technology dispenses liquid binding agent on powder to form a 2D pattern on a layer, and the subsequent layers are stacked to build a physical article. The binder jetting process uses two materials: a powder-based material and a binder. The binder acts as an adhesive between powder layers. The binder is usually in liquid form and the build material in powder form. This technology can be adopted for any powder and offers high production rates. However, the final products are often green and needs post-sintering to attain required mechanical strength.

As shown in Figure 2.9(a), the process starts when a layer of the powder feedstock is placed on the build platform (right-handed) by the powder roller. The inkjet print-head moves over the build platform in the X and Y axes and deposit a pre-calculated amount of binding agent, according to

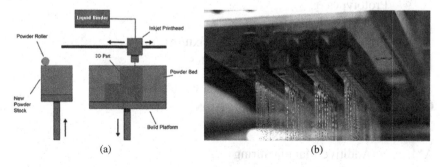

(a)   (b)

Figure 2.9   (a) Schematic representation of BJP (adopted from Engineers Garage) and (b) pictorial view of binder jets (adopted from ExOne).

the layer cross-section. As the given layer is completed, the build platform lowers down (along Z axis) by an amount equal to the selected layer thickness. On the other hand, the supply platform lifts up to provide fresh loose powder from the powder roller in the build area. Again, the print-head moves and provides binding agent droplets on the new loose powder. The process, as usual, follows until the required layer has been bonded.

This process also doesn't require support structures and the loose powder of the build platform provides the required support to the internal cavities and over-hanged structures. After the part printing completes, curing and sintering is performed. Depending on the applications, a secondary material can also be incorporated. BJP has the ability to print very large objects, such as room-sized architectural structures. BJP has received attention due to its high speed and low processing cost.

The various benefits, limitations, and applications of BJP are given below:

- Benefits of BJP
  - Low cost as printing heads are readily available
  - Multi-coloured parts can be manufactured
  - Can be used for metals, polymers, and ceramics
  - The process is faster than other polymer AM systems
  - Can be adopted for larger parts
- Limitations of BJP
  - Poor mechanical properties
  - Post-processing is generally required which may cause damages
  - Sometimes is unsuitable for structural parts as binder is being used
- Applications of BJP
  - Prototypes
  - Casting patterns
  - Bones, skulls, and orthopaedic fixtures
  - Electronic and electrical substrates

Glossary of Key Terminologies

3D: Three-dimensional
AM: Additive Manufacturing

BJP: Binder Jet Printing
CAD: Computer-Aided Design
CO_2: Carbon Dioxide
CT: Computer Tomography
FDM: Fused Deposition Modelling
IJP: Inkjet Printing
LOM: Laminated Object Manufacturing
MRI: Magnetic Resonance Imaging
Nd:YAG: Neodymium-Doped Yttrium Aluminium Garnet
OLED: Organic Light Emitting Diode
SGC: Solid Ground Curing
SLA: Stereolithography
SLS: Selective Laser Sintering

Exercise Questions

Q1: What are the different types of polymer-based AM processes?

Q2: Differentiate between inkjet printing and binder-jet printing.

Q3: Using a suitable diagram, explain the working procedure of an SLS printer.

Q4: What is SLA? Give the various benefits, limitations, and applications of this process.

Q5: Discuss the SGC process in detail.

Q6: What are the different types of polymer materials available for AM technologies? Discuss.

Q7: What is the FDM process? Explain in detail.

Q8: Write a short note on the inkjet printing process. Also draw a neat sketch of the printing system and label various components.

References

1. https://www.globenewswire.com/news-release/2020/03/09/1997179/0/en/Polymer-Additive-Manufacturing-AM-Markets-Applications-2020-2029-Polymer-3D-Printing-Expected-to-Generate-as-Much-as-11-7-Billion-in-2020.html

2. Stansbury JW, Idacavage MJ. 3D printing with polymers: Challenges among expanding options and opportunities. *Dental Materials*. 2016 Jan 1;32(1):54–64.

3. Azad MA, Olawuni D, Kimbell G, Badruddoza AZ, Hossain M, Sultana T. Polymers for extrusion-based 3D printing of pharmaceuticals: A holistic materials–process perspective. *Pharmaceutics*. 2020 Feb;12(2):124.

4. Black HT, Celina MC, McElhanon JR. Additive Manufacturing of Polymers: Materials Opportunities and Emerging Applications. Sandia National Lab. (SNL-NM), Albuquerque, NM (United States); 2016 Jul 1.

5. Melchels FP, Feijen J, Grijpma DW. A review on stereolithography and its applications in biomedical engineering. *Biomaterials*. 2010 Aug 1;31(24):6121–6130.

6. Schmidleithner C, Kalaskar DM. Stereolithography. IntechOpen.

7. Napadensky E. Inkjet 3D printing. The chemistry of inkjet inks. 2010:255–267.

8. Ligon SC, Liska R, Stampfl J, Gurr M, Mülhaupt R. Polymers for 3D printing and customized additive manufacturing. *Chemical Reviews*. 2017 Aug 9;117(15):10212–10290.

9. Mahindru DV, Priyanka Mahendru SR, Tewari Ganj L. Review of rapid prototyping-technology for the future. *Global Journal of Computer Science and Technology*. 2013 May 31.

10. Lee H, Lim CH, Low MJ, Tham N, Murukeshan VM, Kim YJ. Lasers in additive manufacturing: A review. *International Journal of Precision Engineering and Manufacturing-Green Technology*. 2017 Jul 1;4(3):307–322.

11. Ziaee M, Crane NB. Binder jetting: A review of process, materials, and methods. *Additive Manufacturing*. 2019 Jun 22.

Chapter 3

Metal-Based Additive Manufacturing System

This chapter provides fundamental insights into different types of metal-based Additive Manufacturing (AM) technologies including their processes, feedstock materials, and potential applications. Some versatile multi-material AM systems suitable to manufacture both metallic and polymeric feedstock materials, for example laminated object manufacturing (LOM), selective laser sintering (SLS), and binder-jet printing (BJP), were already covered in Chapter 2 and thus have not been included in Chapter 3.

3.1. Metals and Alloys

Metals and alloys represent the most common and extensively diverse class of engineering materials which have been used for producing objects with higher mechanical strength, structural stability, electrical and thermal conductivity, anti-corrosion and wear resistance as compared to any other available class of materials. The main difference between a metal and an alloy is in terms of their metallurgical structure. The metal is made of a singular element, and forms a homogeneous chemical mixture, for example iron (Fe). Metals are considered as chemical solutions rather than compounds and possess the highest possible purity levels. However, an alloy is a mixture of multiple elements of different metals or metals and non-metals. The metallurgical structure of an alloy could be homogeneous or heterogeneous depending on the type and efficiency of processing. For example, adding metal "Fe" and non-metal "carbon" forms steel alloys, whereas adding metal "copper" and another metal "tin" forms bronze alloys. Alloys, therefore, are considered as the solution mixtures of multiple metallic or metallic/non-metallic phases. Indeed, there is a distinct difference between the various properties of metals and alloys. Alloys obtain higher performance characteristics, in terms of the aforementioned properties, than metals.

Nowadays, most engineering applications employ alloy-based materials to manufacture products with greater robustness. The raw metals and alloys can be available in liquid solutions, billets, plates, sections and ingots, powder particles, and even thin sheets. However, most metals and alloys are produced using the ores extracted from the earth's crust through a metal casting process. This is the primary production process for metals and alloys where the ores contain metals along with flux and other additives to remove unwanted wastes exiting along with the ores. In the case of alloys, the specifically required elemental structure is defined during casting by adding other elements in calculated proportions. After casting, there could be a wide range of processes including forging and rolling for producing raw feedstock for different manufacturing processes.

In AM processes, the raw metallic and non-metallic materials required for manufacturing must be either in the form of powder particles, having highly controlled dimensional size and structural shapes, or laminate sheets of defined thickness. For instance, for producing metal powders, various commercial processes (such as solid-state reduction, atomization, electrolysis, and others) exist which further add onto the cost to the raw material. This is, indeed, the prime reason for expensiveness of the raw materials used in metal AM technologies. Following are some of the key terminologies often used in context of metal AM.

- *Ferrous materials:* Materials consisting of a significant proportion of iron are referred to as ferrous materials.
- *Non-ferrous materials:* Those materials which have no proportion of iron are known as non-ferrous materials.

- *Melting temperature:* This is the temperature at which metals and alloys starts to exhibit fluidic properties due to the transformation from solid to liquid state.
- *Sintering:* A process by which the adjoining metallic or alloy particles are fused together by using externally supplied thermal energy.
- *Atomic structure:* The fashion or trend followed by the atoms of a molecule.
- *Grain size:* This is the size of individual grains of sediment. Generally, it is referred to as average grain size.
- *Grain structure:* It is the distribution pattern of the grains in a unit square area.
- *Microscopy:* The technique used for observing the micro and nano details of a finished surface of a metal or alloy.
- *Topography:* It is the surface characteristics of a manufactured or existing object.
- *Post-processing:* This includes the additional processes required for enhancing the aesthetic, geometrical, and surface topography of as-manufactured objects.

3.2. Metal Feedstock

The metal feedstock systems for existing AM have given new life to end-user applications. Despite the fact that metal AM techniques arrived around a decade later than polymer AM techniques, their usage in industrial sectors is far more extensive. This is because of their focus beyond just prototyping. All the credit goes to the wide range of engineering metals and alloys available for AM processing. In comparison to the products made through conventional manufacturing methods, metal AM components have shown stronger potential and found to be more reliable than their counterparts. In today's manufacturing era, many big commercial giants such as General Electric, BMW, SpaceX, and NASA have been taking advantage of metal AM processes for catalysing the efficiency of their production systems and also products. Furthermore, the availability of biomedical grade metal feedstock for AM has brought about tremendous change in the medical sector. The hybridizations of system flexibility, ease of production, and efficient and practical metal feedstock enabled the development of shapes, designs, and applications which were earlier impossible. Experts assert that metal AM systems must not be criticized for their higher capital and operational costs, as unlike desktop printing via polymer AM technologies these are made to support realistic engineering applications by providing the required end-user products with necessary features. Indeed, ongoing trends suggest both hobbyists and industrial systems will drive evolution of the AM market presently. However, as per the predicted future market of this technology, the metal-based system will rise exponentially.

Today, a variety of metallic materials, such as stainless steel, titanium alloys, aluminium alloys, cobalt-chromium alloys, Inconel alloys, and some precious metals, is readily available for converting the three-dimensional (3D) computer-aided design (CAD) model into a mature engineering product. Apart from this, the development of new and more efficient metallic feedstock systems is considered as one of the most focused areas of research in this particular field.

Generally, both pure metals and alloys are being employed in AM techniques; however, material-based requisites make them unsuitable for the whole variety of metal AM systems. To add further, not all metals and alloys can be used in AM systems. Qualification of an available metal powder for a specific purpose is usually mentioned in the specifications sheets that come with the feedstock. It is believed that the continuous expansions in AM systems will soon overcome the rigidity of raw material for multiple system utility. The most valuable applications of metal AM systems belong to automotive, aerospace, and orthopaedic productions. Following the unique "Additive" principle, the systems build the objects by adding the metal/alloy layer-upon-layer; the different procedures adopted are given below:

- *Binding jetting:* Similar to polymer manufacturing, the binding agent is placed in the form of tiny drops onto solid powder particles to impart the green-strength. The post-sintering process imparts the desired mechanical strength to the object.
- *Sintering:* The metallic powder particles are heated and fused together, by using a high intensity laser or electron beam.

- *Melting:* In place of sintering, the powder particles are melted to fuse together with stronger bonding. This needs much higher thermal energy than sintering.
- *Stacking:* Similar to the stacking of polymer sheets, as detailed in Chapter 2, the raw material is in the form of adhesive coated metallic sheets. They are joined with one another through heating and the final shape is given by a laser-based cutting source.

The availability of metal/alloy feedstock systems for this category of AM technology is listed in Table 3.1 for better understanding.

3.3. Metal-Based AM Systems

The variety of metal AM processes can be categorized on the basis of the type of heat source, such as electron beam and laser, and also on the basis of the supply of raw material. Indeed, the rate at which heated metallic feedstock cool to room temperature conditions affects the geometrical shrinkage and other obtained mechanical characteristics [1]. The metal AM system, in comparison to other AM classes, possesses the following key benefits [2]:

- Advanced metallic materials processed through AM technologies are used for aerospace and automobile components, for example structural components, engine valves, and turbocharger turbines.
- The manufactured products possess optimum strength-to-weight ratios.
- The metal AM technology offers the ability to produce customized moulds with optimized cooling channels.
- It provides a competitive alternative for dental implants that rivals traditional casting and milling methods.
- It produces parts with better optical, thermal, electrochemical, and mechanical properties.

The following sub-section provides technical insights into the various metal AM systems, including their key benefits, limitations, and potential applications.

Table 3.1   List of commercially available metal/alloy feedstock for AM systems.

| State | Material | Sintering | | Melting | | | Stacking | Binding |
		SLS	DMLS	LBM	EBM	LENS	LOM	BJP
Powder	Stainless steel	✓	✓	✓	✓			✓
	Tool steel	✓	✓	✓	✓	✓		✓
	Aluminium alloy	✓	✓	✓	✓			
	Cobalt-chromium alloy	✓		✓	✓	✓		✓
	Cobalt							
	Nickel alloy	✓	✓	✓		✓		✓
	Pure titanium					✓		✓
	Titanium alloy	✓	✓	✓	✓	✓		
	Tungsten							✓
	Copper					✓		✓
	Chromium zirconium copper alloy		✓	✓				
	Stainless steel/bronze composite							✓
	Tungsten/bronze composite							✓
	Carbides of tungsten, chromium, and titanium					✓		
	Refractories (vanadium, molybdenum, and niobium)					✓		
	Precious metals (gold and silver)		✓	✓	✓			
Laminate sheet	Tool steel						✓	
	Aluminium						✓	

Note: SLS, DMLS, LBM, EBM, LENS, LOM, and BJP refer to selective laser sintering, direct metal laser sintering, laser beam melting, laser engineered net shaping, laminated object manufacturing, and binder-jet printing, respectively.

3.3.1. *Electron Beam Melting*

Electron beam melting (EBM), sometimes also referred to as selective EBM, is a metal AM technology that uses an electron beam to melt layers of metal powder. This technology was introduced in 1997 by a Swedish company, Arcam, as an ideal approach for fabricating lightweight, durable, and dense products for aerospace, medical and defense industries. The schematic illustration of the EBM process is given in Figure 3.1(a).

Like any other AM technology, EBM is a direct CAD-to-metal fabrication process that can manufacture fully dense solid parts by selectively melting the feedstock metallic powder particles, layer by layer, using a high intensity electron beam. In this process, the standard tesselation language (*.STL*) data of the designed object has been fabricated with the utmost level of accuracy. The pre-heating of powder particles is carried out through a series of defocused beam passages at high power and speed. In this system, the availability of high energy electron beams ensures complete melting of the powder particles and metallurgical bonding between layers which results in the complete part. Along with this, the parts are generally fabricated in a vacuum tank to assure processing of products without causing any type of oxidation and other chemical reactions with atmospheric species. Furthermore, the residual stress caused by straining of the powder particles when they are bonded is minimal because of the in-process vacuum processing. Upon building the parts, they are cleaned to remove loose powder particles lodged within the porous structure. For this, a blast of high pressure airstream is enough. The mechanical properties of EBM build parts are on par with those of conventional manufacturing processes [3].

As Figure 3.1(a) shows, EBM consists of a tungsten filament, a powder container, a spreader, and a build table. The tungsten filament produces a highly focused beam of electrons that passes out from the head. There are two magnetic fields, focus and defluction coils, which help the produced electron beam to follow a defined path towards the powder container. Upon striking the loose powder particles, the kinetic energy of these electrons is converted into thermal energy that causes the fusion of the metallic powder particles with one another. Apart from producing the desired thermal energy, the electron beam also scans the powder particles

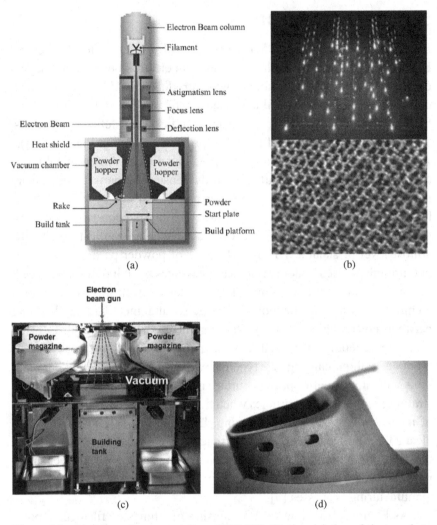

Figure 3.1   (a) Schematic representation of EBM [4], (b) pictorial view of electron shots (above) and as-produced porous architecture (below) [5], (c) in-build view of EBM [6], and (d) as-manufactured EBM product (adopted from CRP Meccanica).

as per the instructions given by the CAD file. The electron gun, in a scanning electron microscope, is capable of operating at a power of 60 kW to generate a focused beam of energy density more than 100 kW/cm^2. For building the objects, a powder layer of thickness 100 μm is spread over the table. The ongoing powder is supplied from hoppers, placed by the rake, from the inside of the build tank. Once the fused powder particles solidify, the build table lowers down and the spreader places another layer of loose powder particles over the previously fused particles.

Then, the electron beam again scans the powder particles and imparts the electrons with high momentum and kinetic energy. This cyclic process continues until the part is completed. The whole process is taking place in the bed of metallic powder particles which provides supports to the over-hanged structures. Therefore, additional solid structures, like in the case of fused deposition modelling, are not required in EBM. To note, the completed part is allowed to cool inside the tank that is filled with helium to assist cooling. Since electrons are radiated in EBM, the observation process is not accessible through a leaded-glass viewport. Further, Figure 3.1(b) shows a pictorial view of the electron gun shots on the build table as well as as-produced porous architecture for biomedical implants. Figure 3.1(c) and 3.1(d) show the in-build setup of EBM system and the as-manufactured product, respectively.

In EBM, the quantity of heat transferred to the powder particles depends on beam power, beam spot size, and scan speed. In addition to this, pre-heating of the powder particles, using a defocused beam, partially sinters the powder and reduces gap between the particles. To further note, the interaction between the electron beam and powder results in [4]:

- Spread of particles
- Sintering of particles
- Melting of particles
- Evaporation of some alloying elements

The various benefits, limitations, and applications of EBM are given below:

- Benefits of EBM
  - EBM's beam is highly powerful, ultimately resulting in faster printing speeds
  - Quality of the parts is very high, comparable to the most precise traditional manufacturing methods
  - The EBM parts have high density

- o   Pre-heating minimises residual stresses
- o   No support structures required
- o   It offers minimal waste as most of the unused powder can be recycled
- Limitations of EBM
  - o   The thicker layers built often result in rough surfaces and, therefore, additional post-processing is required
  - o   The choice of materials is limited as it needs high-quality and expensive materials
  - o   The capital cost of EBM is also very high
  - o   The necessary post-processing may cause damages
- Applications of EBM
  - o   Medical products
  - o   Aerospace and automobile components
  - o   Military devices
  - o   Heat exchangers
  - o   Power unit components

3.3.2. *Laser Beam Melting*

Laser beam melting (LBM), also known as selective laser melting (SLM), is a process similar to the selective laser sintering (SLS) detailed in Chapter 2. However, the intensity of the laser energy is comparatively much higher that makes it suitable for direct melting of metallic powder materials and, therefore, eliminating the post-heating assisted curing of green products as is the case in SLS. In comparison to SLS, this technique provides a lot of benefits including near-net shape production, high material utilization rate, no geometric constraints, and no additional post-processing. The SLM technology was developed by Dr. Fockele and Dr. Schwarze of F&S Stereolithographietechnik GmbH, in collaboration with Fraunhofer ILT. It also starts from a CAD design and builds the metallic part by selectively fusing the regions of metallic powder, layer by layer. The important system processing features are described below:

- It uses a protective atmosphere of argon or nitrogen to minimize the possible surface oxidation and hydrogen pickup

- The thin layer of metal powder is provided on the build platform by the powder depositor depending on the thickness of the slice
- The system scans the powder bed in a pre-defined pattern to produce layer-wise shapes
- The build platform lowers by a pre-determined height, depending on the layer thickness

Figure 3.2(a) shows the schematic description of the SLM process. The process starts with the deposition of a thin layer of powder thickness ranging from 50–75 μm on the build platform. After the powder is laid, a high energy density laser is used to melt and fuse selected areas according to the processed data (refer to Figure 3.2(b)). This creates the molten pool. Furthermore, some of the variants of SLM machines are capable of providing pre-heating to either the substrate plate or the entire building chamber. This feature ensures a good balance between fine resolution and good powder flowability. Then the platform drops down by a single-layer thickness in the Z axis and a fresh powder layer is deposited by the powder depositor arm. The process is repeated until the entire object is completed. Loose powder particles are removed once the fully dense part is complete.

As mentioned above, the whole process is carried out in an inert gas atmosphere to protect the object from any contamination. Processing inputs, such as laser power, scanning speed, hatch spacing, and layer thickness, should be adjusted in accordance to the thermal characteristics of the feedstock material so that a single scan of the laser should be sufficient to produce the melt. Once the laser scanning process is completed, loose powders are removed from the building chamber and the component can be separated from the substrate plate manually. Other than the data preparation and removal of fabricated component from the building platform, the entire process is automated. The post-manufacture heat treatment and material infiltration are often essential to improve the part quality of SLS components. However, in the case of SLM, a complete melting of the powder is achieved by the use of a high-intensity laser without any binder materials, therefore eliminating the need for the additional processes. The currently available commercial SLM technologies provide improvements in product quality, processing time, and

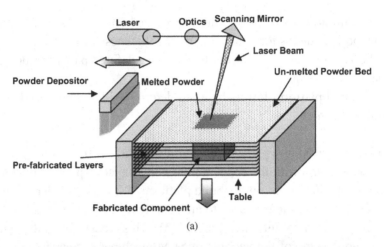

(a)

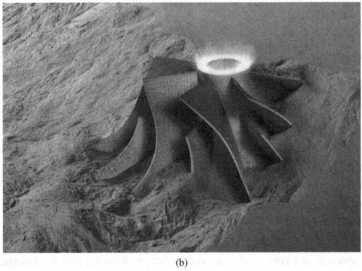

(b)

Figure 3.2   (a) Schematic of SLM process [9] and (b) pictorial view of laser strike on metallic powder (adopted from Manufacturing Guide).

manufacturing reliability compared to SLS [7]. Considering the other side of the coin, the SLM technology has to deal with some downsides, including thermal gradients and rapid solidifications, which cause thermal stresses and unbalanced phase equilibriums. These affect the metallurgical, topographic, mechanical properties, and residual stresses [8]. The various benefits, limitations, and applications of SLM are given below:

- Benefits of SLM
  - It produces highly dense metal parts
  - Speed, accuracy, and ability to create fine details are excellent
  - The fabricated objects are strong and durable
  - It can be hybridized with computerized numerical control (CNC) machines to optimize the smoothness
- Limitations of SLM
  - Extensive cracking due to huge thermal fluctuations
  - Limited flowability after melting and inability to form a compact layer
  - The intrinsic porosity severely reduces the mechanical properties of the part
  - High pre-heating of chamber is required to minimize the temperature gradient
- Applications of SLM
  - Functional prototypes for end-user applications
  - Medical devices
  - Aerospace and automotive components
  - Mechanical fixtures and tools

3.3.3. *Direct Metal Laser Sintering*

Direct metal laser sintering (DMLS) is a sister technology of the SLM process. Similarly, this technology can be considered as an advancement of powder metallurgy (PM). The main difference between SLM and DMLS is the temperature used to fuse the metal powder. In the case of SLM, the thermal energy is required to fully melt the solid powder particles, while in DMLS the intensity of energy required does not melt the metal powder. The sintering of the heated particles is enough to develop the welding between the particles. Owing to this, in DMLS alloys are preferred to pure metals. The working material, like SLM and EBM, is fine metal powder, ranging 20–40 μm. The combination of fine metallic powder particles and better resolution of DMLS allows the fabrication of very precise and near-to-net shape products. The printing layer height in this process is equivalent to the maximum size of the powder particles. The improved mechanical properties and accuracy of DMLS parts have opened up new possibilities for rapid tooling and manufacturing. DMLS can be used for two main applications [10]:

- Tool inserts for the injection moulding of plastics, casting of aluminium, and direct production of steel components.
- Geometries, including complicated channels of automotive, electronics, and household appliances, that otherwise are difficult to produce by conventional methods.

Further, the key features of DMLS are given below:

- Laser sintering of a cross-sectional area
- Net-shape process with minimal shrinkage
- Fully automatic and computer controlled
- No polymer binders
- Scanning optics and CO_2 laser

Figure 3.3(a) presents schematic details of the DMLS process. To start the process, the work table is manually filled with finely levelled feedstock powder. Then, the head laser source scans the 2D cross-section that is precisely switched on and off during exposure of designated areas. The absorption of energy by metal powder will generate the cure and sinter of the already solidified areas below. After this, the work table is moved downwards, along the Z axis, and the material container moves up. The work table and material container move equivalent to the height of the layer thickness. The ruler then moves over to place a new loose powder layer over the previously sintered layer. Any unwanted material is then trashed into the rest material container for collection. The collected material could be used again for the next production run. The DMLS process proceeds layer-by-layer until all parts in a job are completed. In a few hours the machine can produce 3D parts with high complexity and accuracy. In addition to this, during building, the sintered parts reach more or less their final properties; however, as per the requirements of the application the necessary tempering or surface treatment operations are still performed. Figure 3.3(b) and 3.3(c) show some of the practical examples of DMLS products.

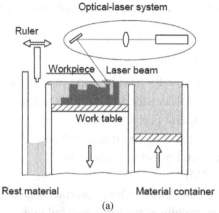

(a)

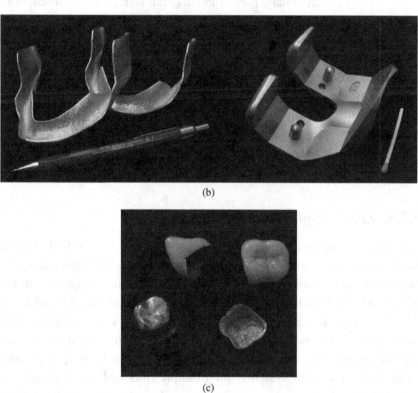

(b)

(c)

Figure 3.3 (a) Schematic representation of DMLS [10], (b) jaw implants and artificial knee joints [11], (c) finished dental restorations [11].

The various benefits, limitations, and applications of DMLS are given below:

- Benefits of DMLS
  - DMLS can be used for a wide range of engineering alloys, including steels, stainless steels, aluminium, titanium, nickel alloys, cobalt chrome, and precious metals
  - It can fabricate strong and functional parts
  - Unsintered metal powder is reusable
- Limitations of DMLS
  - Suitable for only small volume productions
  - The parts are usually porous and may fail under extreme conditions
  - The capital cost of DMLS is very high
- Applications of DMLS
  - Oil and gas industry
  - Automobile and aerospace components
  - Turbine engine components
  - Biomedical implants
  - Regenerative cooled nozzles
  - Bones, skulls, and orthopaedic fixtures

3.3.4. *Laser Engineered Net Shaping*

The laser engineered net shaping (LENS®) apparatus was built by the Sandia National Laboratories as an innovative method to manufacture 3D metallic, ceramic, and composite components as per the designed features of the CAD models. The LENS® technology has demonstrated its potential not only for creating new products but also repairing existing engineering components by following the cladding principle. It is now used for producing cutting-edge models, tooling, and end-user products. It operates like other laser-based AM systems, for example SLS and LBM processes, however it doesn't need a powder bed unlike the other technologies. Furthermore, the laser source in this process can move up and down (along Z axis), while the build platform on the lower side of the apparatus can move along X and Y axes (refer to Figure 3.4(a)). One of the greatest

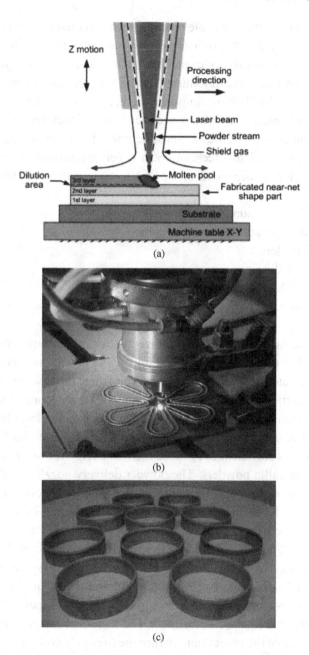

(a)

(b)

(c)

Figure 3.4 (a) Schematic of LENS® [14], (b) pictorial view of LENS® process [13], and (c) thin walled products of LENS® (adopted from Sandia National Laboratories).

features of this technology is the creation of the coordinate in a reasonable and promising manner to produce near-net shapes. Furthermore, this system can be used in various capacities, including rapid prototyping and tooling, and divergent metal joining. The LENS® process is a highly creative innovation that has accomplished criticalness and has just joined manufacturing enterprises. The process can create thick metal parts with good metallurgical properties at sensible rates. However, being a highly sophisticated and efficient technology, it is very expensive. As per Figure 3.4(a), the various components of the LENS® system are given below [12]:

- Nd: YAG laser source
- Self-restrained atmosphere glovebox
- Powder feeders
- Argon shielding gas
- Movable table
- Moving nozzle

The powerful Nd:YAG laser is utilized to cast the fed powder particles, ranging 38–150 μm, onto the build platform. The powder in this system is supplied from the sideways of the feeder unit where the laser energy melts the powder particles and creates a melt pool on the substrate. The molten droplets are then deposited one upon another in a manner described by the CAD model at incremental heights of the nozzle. To note, the fiber laser beam of 2 kW is sufficient to produce thermal energy for most of the high temperature metallic powders. The powder delivery nozzles feed the predetermined amount of powder feedstock into the melt pool on the substrate. After a layer has been formed, the depositing head assembly consisting of focusing lens and powder delivery nozzles moves up by a specified layer thickness and then the second layer is deposited. The procedure is repeated as many times as required to obtain the desired part [13].

LENS® produces a very fine weld bead and the as-produced "heat-affected zone" is smaller and more controlled such that repair of the damaged surfaces can also be performed. The excellent material properties enhance the life of the objects and reduce the life cycle costs. Furthermore, it precisely adds material to the damaged areas with minimal heat effect to enable the manufacture of sensitive thin-wall sections. The LENS®

system is often integrated with conventional processes to create unique hybrid solutions [14]. However, residual stresses due to LENS® processing result due to the uneven heating and cooling cycles.

The various benefits, limitations, and applications of LENS® are given below:

- Benefits of LENS®
  - Fabricates complex products, novel shapes, and hollow sections
  - Reduces the production cost and time-to-market
  - Excellent material properties
  - Capable of manufacturing metallic, ceramic, and composite products
  - Associated waste is minimal
  - Low heat inputs
  - Less distortions
  - Highly effective laser source
  - Near-net shape manufacturing
  - Feasible for hybrid manufacturing
  - Can be used for repair activities
- Limitations of LENS®
  - Thermal distortions
  - Capital cost is high
  - Feedstock systems are very expensive
- Applications of LENS®
  - Aviation
  - Defence
  - Oil and gas industry
  - Tooling, die, and mould
  - Space shuttle components
  - Automotive industry

Glossary of Key Terminologies

3D: Three-dimensional
AM: Additive Manufacturing
BJP: Binder Jet Printing

CAD: Computer-Aided Design
CNC: Computerized Numerical Control
DMLS: Direct Metal Laser Sintering
EBM: Electron Beam Melting
LENS: Laser Engineered Net Shaping
LOM: Laminated Object Manufacturing
RP: Rapid Prototyping
RT: Rapid Tooling
SLM: Selective Laser Melting
SLS: Selective Laser Sintering

Exercise Questions

Q1: What are metallic materials? Explain with examples.

Q2: What are the different types of metallic feedstock available for metal AM systems?

Q3: Explain SLM techniques with the help of a neat diagram.

Q4: With the help of a suitable illustration, define LENS® technology. Also mention its merits and demerits.

Q5: Discuss the EBM process in detail.

Q6: Differentiate between SLM and DMLS processes.

Q7: Explain the working procedures of (a) SLS and (b) SLM.

Q8: Write a short note on BJT. Draw a neat sketch and label its various components.

References

1. Lewandowski JJ, Seifi M. Metal additive manufacturing: A review of mechanical properties. *Annual Review of Materials Research.* 2016 Jul 1;46:151–186.

2. Yakout M, Elbestawi MA, Veldhuis SC. A review of metal additive manufacturing technologies. In Solid State Phenomena. 2018 (Vol. 278, pp. 1–14). Trans Tech Publications Ltd.

3. Parthasarathy J, Starly B, Raman S, Christensen A. Mechanical evaluation of porous titanium (Ti6Al4V) structures with electron beam melting (EBM). *Journal of the Mechanical Behavior of Biomedical Materials.* 2010 Apr 1;3(3):249–259.

4. Galati M, Iuliano L. A literature review of powder-based electron beam melting focusing on numerical simulations. *Additive Manufacturing*. 2018 Jan 1;19:1–20.

5. Sing SL, An J, Yeong WY, Wiria FE. Laser and electron-beam powder-bed additive manufacturing of metallic implants: A review on processes, materials and designs. *Journal of Orthopaedic Research*. 2016 Mar;34(3):369–385.

6. Körner C. Additive manufacturing of metallic components by selective electron beam melting — a review. *International Materials Reviews*. 2016 Jul 3;61(5):361–377.

7. Yap CY, Chua CK, Dong ZL, Liu ZH, Zhang DQ, Loh LE, Sing SL. Review of selective laser melting: Materials and Applications. *Applied Physics Reviews*. 2015 Dec 9;2(4):041101.

8. Manakari V, Parande G, Gupta M. Selective laser melting of magnesium and magnesium alloy powders: A Review. *Metals*. 2017 Jan;7(1):2.

9. Mumtaz KA, Erasenthiran P, Hopkinson N. High density selective laser melting of Waspaloy®. *Journal of Materials Processing Technology*. 2008 Jan 1;195(1–3):77–87.

10. Dolinšek S. Investigation of direct metal laser sintering process. *J. Mech. Eng.* 2004;50:229–238.

11. Kotila J, Syvänen T, Hänninen J, Latikka M, Nyrhilä O. Direct metal laser sintering–new possibilities in biomedical part manufacturing. In Materials Science Forum. 2007 (Vol. 534, pp. 461–464). Trans Tech Publications Ltd.

12. Das M, Balla VK, Kumar TS, Manna I. Fabrication of biomedical implants using laser engineered net shaping (LENS™). *Transactions of the Indian Ceramic Society*. 2013 Sep 1;72(3):169–174.

13. Palčič I, Balažic M, Milfelner M, Buchmeister B. Potential of laser engineered net shaping (LENS) technology. *Materials and Manufacturing Processes*. 2009 May 28;24(7–8):750–753.

14. Ersoy K, Çelik BB. Utilization of additive manufacturing to produce tools. *In Design Engineering and Manufacturing*. 2019 Nov 22. In tech Open.

Chapter 4

Ceramic-Based Additive Manufacturing System

This chapter is dedicated to the different types of ceramic-based Additive Manufacturing (AM) technologies including their processes, various categories of feedstock materials, and potential applications. The versatile multi-material AM systems suitable to manufacture ceramic feedstock materials, for example inkjet printing (IJP), stereolithography (SLA), laser engineered net shaping (LENS®), and binder-jet printing (BJP) have already been covered in previous chapters and are not included in Chapter 4.

4.1. Ceramics

The term 'ceramic' is derived from the Greek word for 'pottery'. A ceramic is an inorganic, non-metallic, often crystalline oxide, nitride or carbide material. Ceramic materials are enormously used for domestic wares, artistic objects, and building elements. The engineering world also takes advantage of more advanced types of ceramics, including glasses, refractory materials, and oxide and carbides of conventional engineering metals (such as tungsten, aluminium, silicon, etc.). The first ever application of man-made ceramic products can be traced back to 24,000 BC when ceramics derived from animal fat and bone were mixed with bone ash and fine clay. The mixtures were fired at a temperature between 500–800°C and shaped in domed kilns, partially dug in the ground.

66

The key properties of ceramic materials are derived from the types of atoms, their bonding, and packing order. It is important to understand that most ceramic materials are made of more than one element. For example, alumina, also known as aluminium oxide or Al_2O_3, is a chemical compound of aluminium and oxygen atoms. Likewise, silicon carbide (SiC) is made of silicon and carbon atoms. Ceramics feature two different types of chemical bonds: covalent and ionic. These bonds in ceramics are much stronger than the metallic bonding found in metals. Owing to this, ceramics exhibit higher mechanical hardness, wear resistance, thermal stability, toughness, and brittleness when compared to metals. There are basically two different categories of ceramic materials, as given below:

- *Natural:* The class of ceramic materials that exists naturally. Diamond, alumina, and SiC are some of the common examples of natural ceramics.
- *Man-made:* This represents another class of ceramic materials which have been synthesized in laboratories. Carbon boron nitride (CBN) is an example of the hardest man-made ceramic material.

All the different types of ceramic materials are brittle and possess very low elastic modulus, making them unsuitable for reliable mechanical applications. Conventional applications of ceramics are for thermal, electrical, and magnetic insulators, optical devices, cutting tools and inserts, abrasive powders and tools, ultrasonic transducers, and semiconductors. Table 4.1 lists different types of ceramic materials and their focused potential applications.

The main compositional classes of engineering ceramics are the oxides, nitrides and carbides, as discussed below:

- *Oxides:* These types of ceramics consist of excess oxygen elements. Al_2O_3 and ZrO_2 are the most commonly used engineering grade oxide ceramics, with alumina being the most used ceramic by far in terms of value and quantity.
- *Nitrides:* These ceramics consist of excess nitrogen elements. Si_3N_4, BN, and CBN are advanced engineering ceramics in this category.

Table 4.1   Ceramic materials and their applications.

Ceramic Material	Abbreviation	Application(s)
Barium titanate	$BaTiO_3$	Electro-mechanical transducers, ceramic capacitors, and data storage devices
Bismuth strontium calcium copper oxide	BSCCO	High temperature superconductor
Boron oxide	B_2O_3	Armour
Boron carbide	B_4C	Abrasive and refractory material
Boron nitride	BN	Lubricant and abrasive
Hydroxyapatite	HA	Biomedical applications
Calcium phosphate	CaP	Biomedical applications
Carbon boron nitride	CBN	Lubricant, abrasive, and cutting tool
Earthenware	—	Domestic ware
Alumina	Al_2O_3	Abrasive, aerospace components and biomedical applications
Ferrite	Fe	Electrical transformers and magnetic core memory
Bio-Glass	—	Biomedical applications
Lead zirconate titanate	PZT	Ultrasonic transducer
Magnesium diboride	MgB_2	Superconductor
Porcelain	—	Household and industrial products
Sialon	Si-Al-O-N	Weld pins
Silicon carbide	SiC	Susceptor for microwaves, abrasive, biomedical applications, refractory material, and aerospace applications
Silicon nitride	Si_3N_4	Abrasive powder
Steatite	—	Electrical insulator
Titanium carbide	TiC	Space shuttle shields and scratchproof watches
Uranium oxide	UO_2	Nuclear reactor fuel
Yttrium barium copper oxide	YBCO	Superconductor
Zinc oxide	ZnO	Semiconductor
Zirconium dioxide	ZrO_2	Cutting materials and biomedical applications
Partially stabilised zirconia	PSZ	Metal forming tools, valves and liners, abrasive slurries, and bearings

- *Carbides:* These consist of excess carbon elements. SiC and B_4C are the extensively used ceramics with thermal conductivity, corrosion resistance, and hardness.

Ceramics are generally mined from rocks and minerals as a starting material and undergo special processing as per the purity, particle size, particle size distribution, and heterogeneity. Chemically or man-made ceramic powders also are used as starting materials. In order to convert the loose ceramic powder particles into the desired shape, water and/or binders are mixed in calculated proportions, followed by a shaping process, including extrusion, slip casting, pressing, tape casting, and injection moulding. After the particles take a green structure, it undergoes a firing- or sintering-based heat treatment to produce a rigid and finished product. Additional processing such as polishing, painting, and machining, also takes place to enhance the aesthetic appearance of the products [1]. Concerning ceramics, the following are some important terminologies:

- *Refractoriness:* It is the property of the ceramic that enables it to withstand high amounts of thermal energy without transforming its physical state. It is generally denoted by the refractory index.
- *Mineral:* It is a naturally occurring solid extracted through geological processes. Minerals have a highly ordered atomic structure and specific physical properties.
- *Crystal structure:* The arrangement of building block elements, such as atoms, ions or molecules, in a crystalline solid. It can be in body centred cubic (BCC), face centred cubic (FCC), or hexagonal close packed (HCP) structure.
- *Binder:* It is an organic or inorganic adhesive material used to bind together the ceramic powder particles.
- *Sintering:* A process by which the adjoining powder particles have been strengthened with the help of thermal energy.
- *Slurry:* A viscous fluid made using ceramic and liquid solutions.
- *Vitrification:* The progressive fusion of a material during the firing process. During vitrification, glassy bond increases and the apparent porosity of the fired product becomes progressively lower.
- *Clay:* A mixture of alumina, silica, and water.

- *Bio-ceramics:* Ceramic materials that can be adopted as biomaterials and can interact with the anatomy of living beings.

4.2. Ceramic Feedstock

Clay is the first ceramic material ever used for manufacturing applications. Since then, a large variety of available ceramic materials have evolved appreciable manufacturing possibilities. The advanced ceramics that exist today have emerged because of the requirement of excellent material systems for demanding applications, such as automotive, aerospace, defence, energy, environmental, and biomedical applications. With continuous technological innovations, the performance and productivity improvement have stimulated the expansion of advanced ceramics. The extensive industrial use of advanced ceramic materials depends on the technology that fabricates near-net-shaped three-dimensional (3D) geometries. The AM here is considered of great importance because it is capable of dealing with a wide variety of ceramics in a rapid and inexpensive manner by eliminating the need of the finishing diamond-based tools usually used in conventional manufacturing [2]. The ceramic feedstock dedicated to AM can be of the following forms:

- Ceramic powder
- Slurries of ceramic powder and binders

The ceramic composites, termed as *"Cermets"*, consist of metallic binders and ceramic reinforcements. These are the advanced materials possessing high hardness and toughness. Some of the common examples of cermets are combinations of carbides, nitrides, oxides, and carbonitrides of titanium, molybdenum, tungsten, tantalum, niobium, and vanadium with nickel, cobalt, and molybdenum alloys as metallic binder. The manufacturing of ceramic products through AM reduces the requirement of dedicated tooling for ceramic parts fabrication. The processing of ceramic materials through AM is very delicate as physical properties of ceramics can only be achieved through careful heat treatment, i.e. sintering. Sintering of ceramics is a delicate step as it ultimately defines the microstructure of the material and geometrical integrities. Therefore, binder-based AM technologies are highly preferred as convenient and cost-effective means of ceramic processing.

Processing principles of ceramic AM technologies are given below:

- *Binding jetting:* Similar to polymer manufacturing, the binding agent is placed in the form of tiny drops onto solid powder particles to impart the green-strength. The post-sintering process imparts the desired glossification and strength to the object.
- *Fusion:* The ceramic powder particles are heated and fused together, by using a high intensity laser beam.
- *Lithography:* The binder ceramic-based liquid slurry is cured using ultraviolet laser source.

The availability of ceramic feedstock systems for this category of AM technology is listed in Table 4.2. Further, Figure 4.1 shows as-manufactured ceramic components for a wide range of applications.

4.3. Ceramic-Based AM Systems

It is a well-known fact that AM technologies can be successfully applied to the manufacture of ceramic-based structures for biological, thermal, and lightweight structural applications. It is crucial to understand that the success of any AM technology, for ceramic manufacture, is dependent not only on the quality of the ceramic feedstock but also on the ability of the AM systems to convert the raw material into precise geometrical structures with required mechanical strength. A major part of the AM of ceramics is dedicated to building porous structures, mainly because of the following two reasons:

- It enables the production of complex architectures which is not achievable by any other technology. In cases of AM, it is not competing with traditional manufacturing, but with itself so that new windows for technological solutions could be opened.
- Most of the ceramic AM technologies are intrinsically suitable for creating fine filigree structures. However, these are not suitable for scaling up to large bulk parts as well as fine geometries [4].

Table 4.2   List of commercially available ceramic feedstock for AM systems.

State	Material	Binding IJP	Binding BJP	Binding Z-Corp	Fusing LENS®	Lithography SLA
Powder	Aluminium nitride					✓
	Silicon nitride					✓
	Silicore					✓
	Zirconia					✓
	Alumina					✓
	Hydroxyapatite					✓
	Silica					✓
	Cordierite					✓
	Alumina toughened zirconia					✓
	Refractories (vanadium, molybdenum, and niobium)				✓	
	Composites (titanium carbide, tungsten carbide, and chromium carbide)				✓	
Slurries	Silica	✓	✓	✓		
	Silicon carbide	✓	✓			
	Boron carbide	✓	✓			
	Alumina	✓	✓	✓		
	Plaster				✓	
	Zirconia	✓	✓	✓		

Note: IJP, BJP, LENS®, and SLA refer to inkjet printing, binder-jet printing, laser engineered net shaping, and stereolithography, respectively.

Slurry-based methods (such as IJP, BJP, SLA, etc.) have shown more promises in controllable feature resolution and surface finish with desirable mechanical performance. However, they can't compete with the LENS® technology in terms of versatility. Selective laser sintering (SLS) also has the potential of generating temperatures high enough for the initiation of sintering processes; however, it is unable to attain a local densification of ceramic powders. The ceramics are highly tolerant to temperature gradients and SLS doesn't provide final physical properties. AM

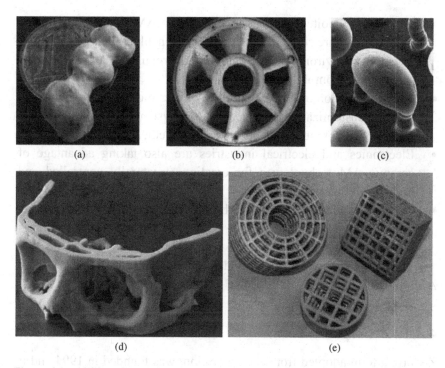

Figure 4.1 AM-based ceramic structures. (a) dental restoration, (b) impeller, (c) micropillar array, (d) bone implant, and (e) macro-cellular structures with variable ligament lengths (adopted from [2]).

technologies can at least be well inserted into the processing route of ceramic materials. In this context, the present communication will discuss approaches for the successful combination of specific features of AM technologies with those of ceramic manufacturing [3]. Despite great progress made in the selection of usable ceramic materials, the optimisation of processing parameters and post-processing is a big obstacle. Furthermore, industrial production at a massive scale is very challenging and components larger than a metre remain difficult to produce with AM owing to the high brittleness and low expansion coefficient of ceramics. In order to extend the applicability of AM to ceramics and realise large-scale production of quality components, it is necessary to base manufacturing mechanism to allow near-net shape production in less time [5]. As of now, ceramic AM systems are benefitting from the following applications [6]:

- The most exploited application of ceramic AM is attributed to the high temperature strength of the products, enabling their practical use in extreme environments of engine and propulsion components for aerospace, automobile and energy.
- The combination of biocompatibility of ceramics and extremely high level of customization provided by AM systems strongly advocates for their employment in dentistry and biomedical applications.
- Electronics and electrical industries are also taking advantage of ceramic AM technologies for creating structurally controlled and high-resolution products.
- Ceramic bioglass products have been used for producing high quality optics through AM.

The following sub-section provides technical insights into the Z-Corp AM systems.

4.3.1. *Z-Corp*

Z-Corp, a term adopted from Z Corporation, was founded in 1994 and is actually based on a technology developed in 1993 at MIT, USA under the direction of Prof. Sachs. Though the processing of Z-Corp may be related to IJP, it is distinct from other AM systems. It has, indeed, good resolution, fast processing, and versatility. The Z-Corp process is similar to the two-dimensional (2D) desktop printers, with a key difference being that additional Z axis movement is given to the build platform. Z-Corp adopts the three-dimensional (3D) printing procedure and benefits rapid proto-typing and tooling (RPT) applications. Owing to this, it was the third best sold printer a decade ago. Despite the various factors involved, the Z-Corp system is able to build parts faster than fused deposition modelling. The most commonly used powder system results in parts that are attractive, aesthetically, but brittle. The fundamental constructional features of Z-Corp manufaturing are:

- There are two chambers, namely feed chamber (on the left) filled with the powder and build chamber (on the right), where the actual part is made.

- The build chamber moves downwards and the feed chamber moves upwards during the building process. Therefore, the build chamber possesses motion along Z axis.
- A powder spread roller is used to place fine layers of powder on the build table.
- An inkjet print-head is mounted on top of the system, to deposit binding liquid/adhesive; it can move along X and Y axes.

Figure 4.2(a) describes the schematic illustration of the Z-Corp process. The processing of Z-Corp starts with a computer-aided design (CAD) model that is converted into standard tesselation language (.*STL*) file before transfering to the system. Then, Z-Corp slices the .*STL* file and generates the tool path according to the reccomended or customized parametric setting. As discussed earlier, the inkjet print-head takes two motions along X and Y axes, while the build chamber takes the third motion along Z axis. Intially, the levelling roller places a thin layer of ceramic powder on the platform of the build chamber. Following this, the inkjet print-head moves and deposits tiny drops of the binder on the pre-decided locations of the thin powder layer. With this, the binder links the ceramic powder particles that upon solidification maintain structural integrity. The loose powder not utilized in the manufacturing operation stays in place and acts as the support. The internal constructional features of Z-Corp can be seen in Figure 4.2(b).

With all these actions, the very first layer of the product has been completed. Afterwards, the powder feed piston lifts up and the build chamber drops down by a distance equal to the thickness of the layer. The levelling roller then supplies another fine layer of the ceramic powder that is placed on top of the previously created layer. Again the inkjet print-head comes into action and deposits tiny drops of the liquid binder onto the defined locations of the loose powder layer. The process repeats a number of times as required for completing the whole object. The part upon completion is taken from the build chamber and is post-sintered to make it stronger and ready for end-user application. Figure 4.2(c) shows the end-user products manufactured by the Z-Corp system.

The Z-Corp system doesn't use any source of laser or electron beam, therefore the running cost is comparatively lower than SLS and LENS®

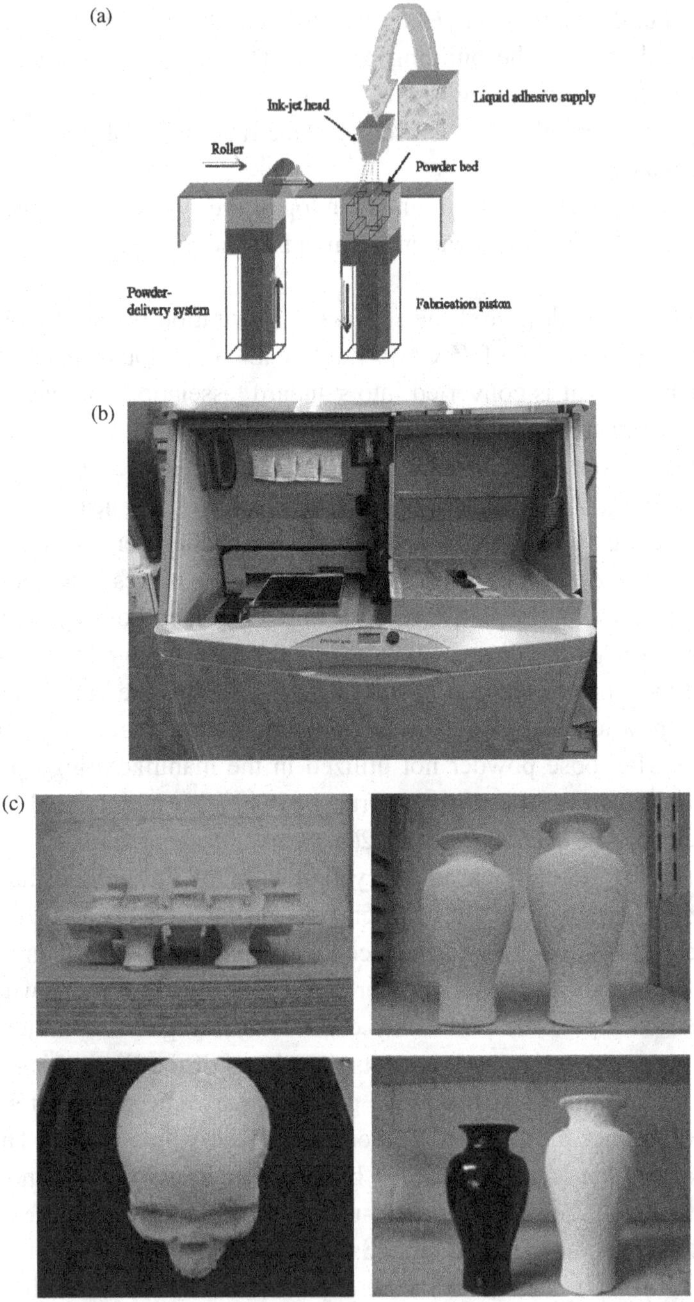

(a)

(b)

(c)

Figure 4.2 (a) Schematic representation of Z-Corp [7], (b) internal view of Z-Corp system (adopted from Instructables), and (c) as-manufactured Z-Corp products [6].

systems. The various benefits, limitations, and applications of this system are given below:

- Benefits of Z-Corp
  - The manufacturing speed of this system is high
  - The building material is not subjected to thermal energies, hence residual stresses developed during fabrication are almost negligible
  - There is no need of additional support structures for the over-hanged sections
  - The process is economical compared to the traditional manufacturing of unique entities
  - The capital and manufacturing cost of the system is low
  - Easy to operate
  - Multiple objects can be made simultaneously
- Limitations of Z-Corp
  - Additional post-processing is required
  - The system has very limited choices of ceramic feedstock material, therefore can be used for only a few dedicated applications
  - Dimensional accuracy of the end products is poor
  - The build size of the part is limited
- Applications of Z-Corp
  - Casting and injection moulds
  - Tableware
  - Dental crowns and tools
  - Architectural components

4.4. Key Issues with Ceramic AM

The applications of ceramic AM are consistently evolving with the invention of new upgraded systems and feedstock materials systems. However, the problems faced by ceramic product manufacturing continue to persist. Mechanical properties including fracture toughness, compression and tensile strength, fatigue life, and durability of the as-manufactured ceramic components present a key threat during extreme service conditions. This is primarily because of the brittleness of the ceramic particles which cause geometrical components to fail immediately without showing any

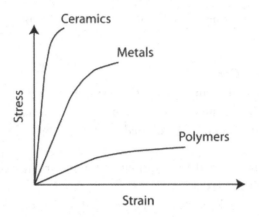

Figure 4.3.   Stress vs. strain plot for ceramics, metals, and polymers (adopted from iMechanica).

elongation or yielding. It means that the load bearing capacity and fatigue life of resulting ceramic products is very nominal as compared to their counterparts. In order to understand it better, let us consider the stress vs. strain plot of a tensile test to compare failures (refer to Figure 4.3). It can be seen that the metallic and polymeric components, owing to their elasticity, undergo substantial elongation followed by necking and then ultimate failure. However, in the case of ceramics, no such characteristics can be seen. Rather, the failure is sudden. On the other hand, the maximum load sustained by ceramics is higher than polymers and metallic components. Therefore, for engineering services, many manufacturing companies are considering ceramic reinforced metallic and polymeric composites so that limtations can be traded off with the individual material systems' strengths.

Poor surface roughness and in-built porosity of ceramic AM are other key issues. Parts are manufactured by binding the ceramic powder particles with each other, therefore the resulting porosity can't be outweighed. By selecting the input process parameters of an AM system, the resulting structural porosity and surface roughness could be significantly controlled, but not outweighed completely. Since the structural porosity reduces the mechanical properties, infiltration of a secondary material to occupy the void spaces between the particles can be considered at an additional cost. Similarly, to make the ceramic components manufactured through AM processes more appealing aesthetically, painting, grinding or filling operations can be adopted by additional spending.

Glossary of Key Terminologies

2D:	Two-dimensional
3D:	Three-dimensional
AM:	Additive Manufacturing
BJP:	Binder Jet Printing
CAD:	Computer-Aided Design
IJP:	Inkjet Printing
LENS:	Laser Engineered Net Shaping
RPT:	Rapid Prototyping and Tooling
SLA:	Stereolithography
SLS:	Selective Laser Sintering
.STL:	Standard Tessellation Language
Z-Corp:	Z Corporation

Exercise Questions

Q1: What are ceramics? Explain in detail.

Q2: Discuss any five key terminologies related to ceramics.

Q3: Write a short note on AM-based ceramic feedstock materials.

Q4: What is Z-Corp? Discuss its various benefits and limitations.

Q5: With the help of a neat diagram, discuss the workings of the Z-Corp system.

Q6: Why is the SLS system not suitable for manufacturing ceramic objects?

Q7: Discuss the difference between natural and man-made ceramics with examples.

References

1. https://depts.washington.edu/matseed/mse_resources/Webpage/Ceramics/ceramicprocessing.htm

2. Travitzky N, Bonet A, Dermeik B, Fey T, Filbert-Demut I, Schlier L, Schlordt T, Greil P. Additive manufacturing of ceramic-based materials. *Advanced Engineering Materials*. 2014 Jun;16(6):729–754.

3. Mühler T, Gomes CM, Heinrich J, Günster J. Slurry-based additive manufacturing of ceramics. *International Journal of Applied Ceramic Technology*. 2015 Jan;12(1):18–25.

4. Zocca A, Colombo P, Gomes CM, Günster J. Additive manufacturing of ceramics: issues, potentialities, and opportunities. *Journal of the American Ceramic Society*. 2015 Jul;98(7):1983–2001.

5. Chen Z, Li Z, Li J, Liu C, Lao C, Fu Y, Liu C, Li Y, Wang P, He Y. 3D printing of ceramics: a review. *Journal of the European Ceramic Society*. 2019 Apr 1;39(4):661–687.

6. Yang L, Miyanaji H. Ceramic additive manufacturing: a review of current status and challenges. *Solid Free. Fabr. 2017 Proc. 28th Annu. Int.* 2017:652–679.

7. Peltola SM, Melchels FP, Grijpma DW, Kellomäki M. A review of rapid prototyping techniques for tissue engineering purposes. *Annals of Medicine*. 2008 Jan 1;40(4):268–280.

Chapter 5

Special Additive Manufacturing System

This chapter presents the special Additive Manufacturing (AM) technologies being used in biomedical, mechanical, furniture, food, and constructional industries.

5.1. Concept of Purpose-Based AM

Thus far, foundation knowledge was provided to understand the basic concepts of different categories of AM technologies, focusing primarily on the type of raw feedstock material required to start manufacturing. The various industrial applications of AM technologies have also been highlighted, wherever possible. As a consequence of increased demand for a greater variety of products, many enterprises are forced to rethink their strategies to offer more product variants as well as for a wider range of applications. In order to satisfy this demand, AM systems are being reconfigured. Generally, AM re-configurability repeatedly changes its capacity, functionality, and cost-effectiveness, in order to meet different market exposures. In this context, a high level of automation might not be the only solution for re-configurability of AM systems, and new innovations are required. Indeed, active research has opened new horizons of AM. The majority of such activities are focused on breaking the limits of existing AM technologies by introducing required changes in the system hardware.

Exemplifying such innovations, special purpose (SP) AM technologies are being developed to provide broader expansion of this unique technology in diverse sectors of importance. All the credits of such breakthroughs are solely dedicated to improvements in the technologies that have enabled its growth beyond just rapid prototyping (RP). Indeed, it is worth mentioning that the huge interest attained by AM is because of the numerous benefits provided relative to conventional manufacturing processes. Being a digital and flexible process, fabrication of the customised products, on demand, is economically more favourable. Although the commercial AM systems are self-capable of addressing a broad range of engineering and scientific applications, the need of SP-AM systems has emerged because of the following key reasons:

- The limited build size of commercial AM systems is unable to produce limitless dimensions as required in the real world.
- The availability of open source or proprietary feedstock materials does not always suit end-user applications.
- The as-produced geometrical and topographic features of AM products are not always better than conventional manufactured products.
- The manufacturing speed and production yield of AM systems are very limited when compared with conventional manufacturing.
- There are restrictions in using novel and efficient feedstock materials other than the system-dedicated type.
- Commercial AM systems are limited in exploring many industrial sectors.

Therefore, the need of the hour acted as the driving force for the development of SP-AM technologies. The original equipment manufacturers (OEMs) are the high-end users of such SP-AM to undertake complex and precision manufacturing of advanced products. To enable such AM systems to operate reliably and produce quality components, the synchronization between feedstock material, process parameter selection, and tool path generation must be robust. This revolution of smart AM systems has just begun but has already triggered the transition to new manufacturing procedures. This can resolve the hike in demand of finished products and strengthen the technology by offering room for reducing associated costs. The SP-AM systems offer many advantages in terms of growth of the technology, expansion of the projected applications, involvement of versatile materials systems and other conventional technologies, and production of physical objects irrespective of their state and applications.

However, the design and manufacturing cost of SP-AM could be high, therefore proper justification of its utilization should be made before any decision to design and manufacture one. Admittedly, the development of a SP-AM system requires appropriate and effective evaluation and identification of the major factors impacting the end-user applications. It is often recommended to alter the design and configuration of existing low cost AM systems than building a completely new technology. Furthermore, technical feasibility analysis is also required for such an activity. Overall performance of upgraded AM systems should be judged on the basis of the following parameters:

- Quality of the end products
- Capital and manufacturing cost incurred
- In-service performance
- Value additions, if any
- Obligations for regulatory standards, if any

The next section discusses the different types of existing SP-AM systems for specialized applications.

5.2. SP-AM Systems

This section provides the details of various SP-AM systems with respect to the end-user industry. For better understanding, the available SP-AM technologies have been classified as given by Figure 5.1.

5.2.1. *Bioprinting*

Before getting into the complex terminologies associated with bioprinting, it is very important to understand the aspects of this particular technology as well as its intent. The emergence of bioprinting technology is

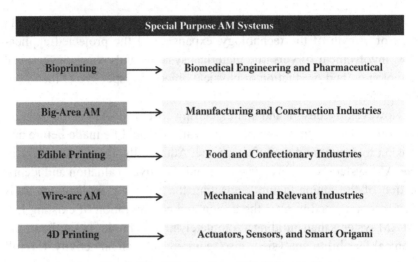

Figure 5.1   Classification of SP-AM systems.

aimed to respond to the rapid growth of patient-specific medical tools and devices, including the skull, skin, medicines, as well as artificial organs (such as heart, kidney, lung, liver, etc.). This technology (refer to Figure 5.2 for schematic illustration) is capable of synthesizing biocompatible living tissues by utilizing the same "additive" principle as well as a source of data, normally the CAD model. Bioprinting is the three-dimensional (3D) manufacturing of biological tissues and organs through the layering of living cells [1].

However, the data model in this case is always preferred to be obtained from computer tomography (CT) or magnetic resonance imaging (MRI) so that the geometrical features of the end-product match the patient specifications. Furthermore, the technology doesn't need any foreign feedstock material as the starting materials; in this case they are isolated from human body cadavers and other living organisms. For example, the starting materials may be solutions of proteins or living human cells in the form of micro-gels. For tissue engineering applications, for instance scaffolds of soft organs like nose and ear, cells are planted on the biodegradable polymer architectures to foster their growth. In the synthesis of micro-gels, chemical or physical compounds are added with the human cells, proteins and/or deoxyribonucleic acid (DNA) so that

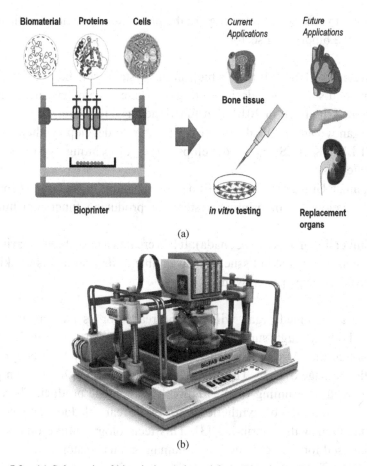

Figure 5.2 (a) Schematic of bioprinting (adopted from Bioprinting Blog) and (b) pictorial view of bioprinting system [2].

Note: In vitro testing means to observe the performance of any medical product/material in a simulated human body environment.

formulated feedstock would be able to obtain the required rheological properties and easily converted into 3D structure. For example, printing the bladder, a simpler organ, consists of two different steps. Initially, a 3D scaffold will be printed to provide a suitable foundation on which the planted human cells can grow. Afterwards, the cell-planted scaffold will be placed in a bioreactor to provide the cells with a suitable environment to grow into an organ. Following are the prominent companies/collaborations in the bioprinting sector:

- Organovo (USA): It makes bioprinting systems suitable for both normal and specifically designed tissues. The company has successfully printed the liver, in 2014, that functioned as a real liver.
- Organovo and L'Oreal (France): They aim to develop synthetic skin.
- CELLINK (USA): It develops both bioprinting systems and materials.
- Spanish Universidad Carlos III de Madrid and BioDan Group (Spain): It developed a bioprinting system for producing functional human skin.
- University of Toronto (Canada): It has created a hand-held bioprinting system to print skin tissue. It is even able to deposit layers of skin to cover and repair deep wounds.

The as-obtained organ will then be ready for its placement in the human body through surgery. The most important point to understand here is that the 3D polymer scaffold will automatically be biodegraded with the passage of time. The overall printing process is highly complex and difficult, consuming weeks to month for a single product. There are real-time examples of synthetic skin, bionic ear, bladder, and cornea manufactured with bioprinting [3]. This technology, unlike others, can only be used for the construction of human organs, anatomical parts and for more efficient drug testing. Although artificial organs have been developed through bioprinting, as of now these have not been used for their end-user applications. The breakthroughs are still undergoing critical research evaluations and only upon successful clinical trials and Food and Drug Administration (FDA, USA) approval, can they be considered for clinical use. There are three central approaches of bioprinting, as discussed below [4]:

- **Bio-mimicry:** The bioprinted architecture should mimic, or behave, in a way similar to the natural biological host tissue in the human body. Therefore, the selection of materials is very critical to suit the essential characteristics desired by the host tissue.

- **Autonomous self-assembly:** This approach adopts the use of embryonic tissues to replicate biological host tissues. It demands that the early cellular components of developing tissues should produce their own extracellular matrix components for cell signalling.
- **Mini-tissues:** This concept is relevant to both of the above approaches. In this, the organs and tissues comprise smaller functional building blocks, mini-tissues, as the smallest structural and functional component.

The material used for bioprinting must possess the following features [4]:

- **Printability:** Ability to get printed/manufactured through bioprinting. It depends on viscosity, gelation methods and rheological properties.
- **Biocompatibility:** Materials should not induce undesirable responses from the host and should contribute actively and controllably to the biological and functional components of the construct.
- **Degradation kinetics:** The rate at which the material degrades should be matched to the ability of the cells to produce their own extracellular matrix.
- **Structural and mechanical properties:** The materials should be chosen based on the required mechanical properties of the construct.
- **Bio-mimicry:** Designed structures should respond similar to the host tissues.

The working principle of the bioprinting apparatus is similar to the common class of AM. Common feedstock materials, mainly consisting of biomaterials, are given in Table 5.1. As of today, there are four main technologies being used for bioprinting as given below:

- **Inkjet Bioprinting:** In this case, the biomaterial is dropped on the build platform, usually a reservoir suitable to protecting biomaterials from undesired physical and chemical changes. The feedstock of inkjet bioprinting is a synthetic bio-ink that passes through the printhead nozzles. The technique can be driven by a thermal source to produce pressure pulses that push droplets out of the nozzle. The average temperature increase only ranges 4–10°C, therefore it doesn't cause any impact on the viability of cells [5]. Piezoelectric pressure

Table 5.1   Common materials for bioprinting (adopted from [6]).

Material	Suitable method				Application(s)
	Inkjet	Extrusion	Laser	SLA	
Nano-cellulose		✓			Wound dressing
Alginate/collegen/gelatin/ gellen		✓			Organs and tissues
M13 phages and alginate		✓			Regenerative tissues
Polyethylene glycol (PEG)- based bio-ink, fibroblasts, sodium alginate, soy agar, polystyrene microbeads, methacrylic anhydride, and 3T3 cells	✓				Soft tissues
Gelatin-based bio-inks		✓			Soft tissues
Poly(ethylene glycol) diacrylate, gelatin methacrylate, eosin Y-based photoinitiator				✓	Micro-scale cell patterning
Hyaluronic acid, acrylated, and pluronic F127		✓			Tissue engineering
MG63 cells, alginate, polycaprolactone			✓		Tissue engineering

pulse could also be used to flow out the bio-ink from the print-head. Figure 5.3(a) shows the schematic of the inkjet bioprinting process.

- **Laser Bioprinting:** Laser bioprinting technology (Figure 5.3(b)), consists of three main parts: (i) pulsed laser source, (ii) ribbon coated with liquid biological materials that are deposited on the metal film, and (iii) receiving substrate. During printing, the lasers expose the ribbon to radiation, allowing the liquid biological materials to vaporize and reach the receiving substrate in droplet form. The receiving substrate contains a biopolymer or cell culture medium to enable the cells to remain adhered [5]. This is a nozzle-free system and has no issues of choking of cells or biomaterials. It is a very costly apparatus.

- **Extrusion Bioprinting:** The extrusion bioprinting system, illustrated in Figure 5.3(c), produces a continuous thin slice of biomaterial and

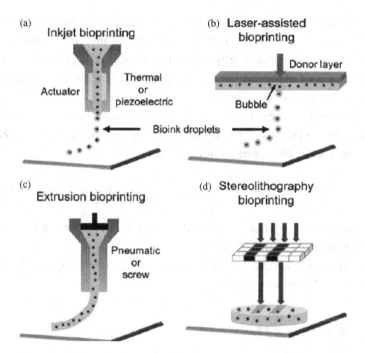

Figure 5.3 Schematic of (a) inkjet-, (b) laser-, (c) extrusion-, and (d) stereolitography-based bioprinting systems [6].

deposits it in a two-dimensional (2D) fashion on a protective reservoir. It uses three main sources of pressure: (i) piston-driven, (ii) pneumatic-driven, and (iii) rotating screw-driven. Extrusion bioprinting is a blend of a fluid-dispensing system and an automated robotic system. Piston-driven and pneumatic-driven deposition normally provide more direct control over the flow of bio-ink from the nozzle. Screw-driven systems give spatial control and are favourable for depositing bio-inks with greater viscosities [5].

- **Stereolithography Bioprinting:** Stereolithography bioprinting, as shown in Figure 5.3(d), is based on polymerization of light-sensitive polymers by precisely controlled light glinted by the digital micromirrors. In comparison with other methods, this technique has high printing quality, speed, and cell viability. However, the UV light source used has been reported to be harmful for DNA-based cells.

A common problem of the structures fabricated by current bioprinting technologies is a lack of mechanical strength and integrity in the printed constructs due to the innate properties of hydrogels. The printed structures should have sufficient mechanical strength to maintain their shape and withstand external stress after implantation. Most hydrogels used in bioprinting systems possess low mechanical properties since bio-ink needs to maintain low viscosity to prevent clogging of the delivery nozzles. Following are the challenges that need attention [7]:

- Less resolution, repeatability, cell viability, and biocompatibility of the bioprinters.
- Insufficient cell density, printability, solidification speed, and mechanical properties of the bio-inks.
- Limited compactness, resolution, accuracy, affordability, and versatility of the bioprinting systems.

5.2.2. *Big-Area AM*

The AM systems discussed so far are facing size constraints which limit the overall dimensions of the build components. This means that while size ranging in inches could be built, the actual applications can be in meters. This gave rise to the Big Area Additive Manufacturing (BAAM) concept that enabled the large-scale manufacturing of engineering components. BAAM was developed by Cincinnati Incorporated and Oak Ridge National Laboratory's Manufacturing Demonstration Facility in the form of a large-scale FDM system. BAAM is capable of producing large complex shapes, while following the similar design constraints applicable to small-scale polymer 3D printers. Additionally, the BAAM system enables the fabrication of large-scale parts because it eliminates the use of an oven as there is no thermal distortion associated with this process [8].

Figure 5.4(a) shows the pictorial description of a commercially available BAAM system. BAAM is an extrusion process utilizing the injection moulding principle. The gantry system can move in the X, Y, and Z directions to build the part. To note, the single screw extruder is capable of delivering 100 lbs/hour of thermoplastic pellet feedstock [9]. Furthermore,

(a)

(b)

Figure 5.4 (a) A pictorial view of the BAAM system [8] and (b) an as-printed wind turbine blade [9].

the gantry system is capable of achieving 200 inch/sec velocities with a position accuracy of 0.002 inches. A carbon fiber reinforced thermoplastic feedstock can also be used for producing the part with adequate resins which increases strength and stiffness [10]. The elimination of the oven decreases the energy intensity required per kilogram of product [11].

However, one of the issues that currently limit the functional use of BAAM printed components is mechanical anisotropy. The strength of the printed parts across successive layers in the build direction (Z-direction) can be significantly lower than the corresponding in-plane strength (X–Y directions).

Specifically, large and complex injection moulds can be created through the system. Figure 5.4(b) shows an as-printed wind turbine blade. In comparison to the small scale FDM printer, the BAAM system produces thermoplastic layers ranging in inches. Apart from the polymer components, there are some examples where the researchers have developed magnetic components [12]. Figure 5.5 shows some large scale AM products [13].

Apart from plastic production, the BAAM concept has also been used for producing large construction and architecture sites. For this, industrial scale binder-jet printers, material extrusion printers (similar to FDM system), and powder bed fusion (similar to SLS system) are adopted . There are numerous examples of AM being used to construct large scale buildings, for instance [14]:

- A two-storey house in China, about 400 m², was built by Beijing-based HuaShang Tengda in 2016.
- An office building in Dubai, UAE, measuring 250 m², was built in 2016 by Chinese construction company Winsun. The building was printed using a $120 \times 40 \times 20$ feet printer.
- The interior of a hotel suite measuring $12.5 \times 10.5 \times 4$ m, in the Philippines, was completed in 2015 by Total Kustom.
- A five-storey apartment building in Suzhou, China, was completed in January 2015 by Winsun.
- In Suzhou, China, a 1100 m² villa was built by Winsun in 2015.
- Children's Castle, Minnesota, USA, was completed in 2014 by Total Kustom.
- A series of 10 houses in Suzhou, China, was built by Winsun in 2014.

Most of the construction-based AM systems deliver the required material by using a gantry system, moving in the Cartesian coordinate

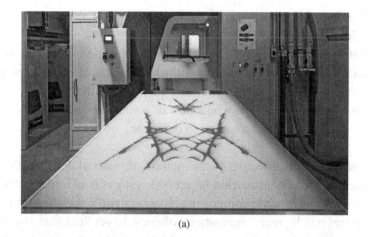

(a)

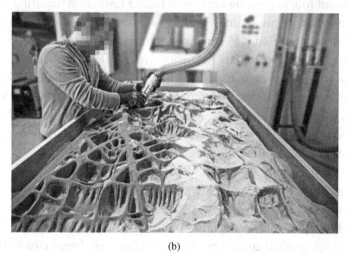

(b)

Figure 5.5 (a) Binder-jet printed component and (b) complex standalone product for construction applications [13].

system, where the nozzle or building platform moves in all the three axes. However, they do have limitations such as:

- **Transportation and installation:** For building a large-scale component, a gantry system must be larger than the component being built.

Therefore, the transportation and installation of the gantry is very difficult.

- **Orthogonal deposition:** The orthogonal deposition is another limitation as the gantry system only allows extrusion of material perpendicular to the build surface, limiting curvature to the horizontal plane.
- **Size of the system:** The whole size of a construction-based AM system is very large and complex to deal with.

Furthermore, irrespective of the method of material delivery, the low manufacture rate of construction-based AM systems is an important drawback. Printing large-scale components takes a significant amount of time and requires much greater deposition volumes. Layer thickness also plays an important role in printing time, with higher resolution requiring thinner layers and more printing time.

Construction-based AM can work with cementitious, metallic, polymer and form-based materials for various value-added applications. There are about fifteen players in the construction AM field (including D-Shape, Contour Crafting, Concrete Printing, 3DCP, XtreeE, BAAM, KamerMaker, Skanska, Arup, Permasteelisa, MX3D, CyBe, Apis-cor, Arup, MX3D, DCP, and FreeFABTM Wax). The industrial applications of these printers are increasing due to the following unique advantages [15]:

- No need for tooling, reducing production time and costs
- Possibility to quickly change designs
- Product optimized for function
- More economical custom product manufacturing (mass customization and mass personalization)
- Potential for simpler supply chain, shorter lead times and lower inventories

D-Shape, Contour Crafting, and Concrete Printing are the most efficient systems for printing cementitious materials; for schematic descriptions refer to Figure 5.6 [16].

- **D-Shape:** The D-Shape printer is a large system that selectively binds sand with a magnesium-based binder in order to create stone-like

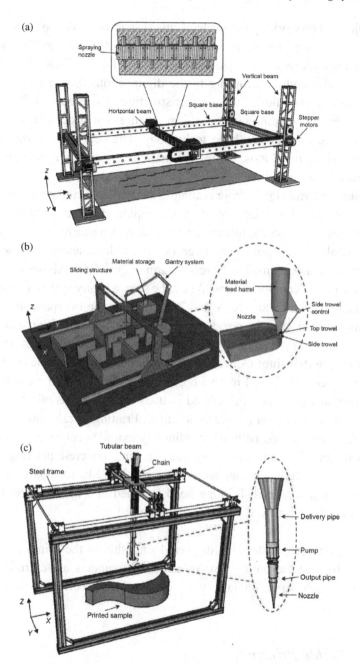

Figure 5.6 Schematic of (a) D-Shape, (b) Contour Crafting, and (c) Cement Printing setups [16].

objects. The working principle of this system is similar to SLS where the sand-based feedstock system is spread out to a desired thickness and the hundreds of nozzles deposit binding liquid to selectively bind the sand together according to the digital prototypes. However, the remaining sand builds the support structure. The process follows the layer-by-layer building mechanism. Figure 5.6(a) depicts the schematic view of the D-Shape technology. The depositing nozzles can move along the X axis, while the square base, containing sand, moves upwards in the Z axis.

- **Contour Crafting:** Contour crafting is a mega-scale automatic construction process that is controlled by a computer. Contour crafting offers better surface quality, higher build speed, and wider range of optional materials. It can print very large objects with dimensions of several meters by using a multi-axis robotic arm. Figure 5.6(b) shows the schematic plot of contour crafting. As can be seen, a gantry system carries the printing nozzle along the X and Z axes, while the two parallel sliding structures carrying the nozzle move along the Y axis. It is an extrusion process which pours the cementitious material on the build platform.

- **Concrete Printing:** Concrete printing is another large scale construction process (refer to Figure 5.6(c)). It is similar to contour crafting to a certain extent as a print head is used for the extrusion of cement mortar, mounted on an overhead crane. Printing nozzle moves along a pre-programmed path and continually extrudes concrete materials. However, in comparison to contour crafting, concrete printing has a smaller resolution of deposition that results in better controlling of complex geometries. Printing head installed in a tubular steel beam can freely move in X, Y and Z directions.

Then, concrete material is delivered smoothly to the printing nozzle with the help of the pump. Finally, concrete filaments are extruded out from the nozzle to continually trace out the cross section of structural components.

5.2.3. *Edible Printing*

One of the most interesting and versatile applications of AM is the processing of delicious edible items with potential benefits. For instance, AM

Figure 5.7 AM of food (adopted from Natural Machines).

of edible items is a healthy and environmentally safe procedure as it helps to convert alternative ingredients to tasty products. Moreover, it has also made cooking interesting as well as easy by opening the door to food customization as per the individual's needs and preferences. As an excellent way of preparing food by utilizing the "additive" principle, it can print the edibles, for instance pizza, by depositing the layers of pizza sauce, chopped vegetables and other toppings, and cheese in a layer-upon-layer fashion. Figure 5.7 shows the pictorial view of the AM of food. It possesses various benefits in comparison to conventional culinary practices as given below:

- **Reliability:** With AM of food, the needed tolerances can be achieved.
- **Speed:** The cooking of the food using AM is convenient for job production. However, this may not be preferable for mass production.
- **Cost:** The AM system for this application is expensive. However, apart from the machinery, the cost of production of the customized food is low.
- **Safety:** The food processed through AM is safe and ready to eat.
- **Creativity:** It is a unique method of converting innovative food ideas into meals, beverages, chocolates, and ice-creams.

- **Sustainability:** AM has the ability to supply an ever-growing world population as compared to traditional food manufacturing systems by minimizing waste.

AM systems can use crystallized fine-grain sugar for manufacturing perfect geometric configurations and high-throughput confectionaries. There are many examples where syringes are being used to dispense chocolate into beautiful patterns. Foodini is employing AM food ingredients for producing dishes. A German nursing home is also using the technology for producing food products such as *Smoothfoods*, consisting of mashed peas, carrot, and broccoli. At the Consumer Electronics Show of Las Vegas in 2014, the Culinary Institute of America showcased the ChefJet system for digital processing of food through AM.

For printing food, heat from the source laser, hot air, heating element, or sprayed binder or solvent can be used for the fusion and joining of the layers. Furthermore, the powder can also be sintered or molten material can be extruded. While AM converts food in a semi-liquid, pure or powder form to edited food products, part of the food after the printing process may require further processing, such as cooking, baking or frying [17]. As there is a growing need to personalize products such as gluten-free, sugar-free, lactose-free, organic, or bio foods, this technology has potential to become mainstream [18]. The following technologies, with schematics given in Figure 5.8, are applicable to food 3D printing:

- **Fused deposition modelling, also referred to as hot-melt extrusion:** The molten or liquid food ingredients, for instance chocolate, is extruded on the platform.
- **Hot-air extrusion:** Hot air is extruded from the nozzle to bake the powder particles of the food items.
- **Selective laser sintering:** In place of the hot air, laser is used for baking the powder particles of the food items. It uses low melting solutions such as sugar, NesQuik, and fat.
- **Binder-jet printing:** The binding particles are deposited from the jet nozzles onto the powdery food items. It uses powders of sugar, starch, corn flour, flavours, and liquid binder.

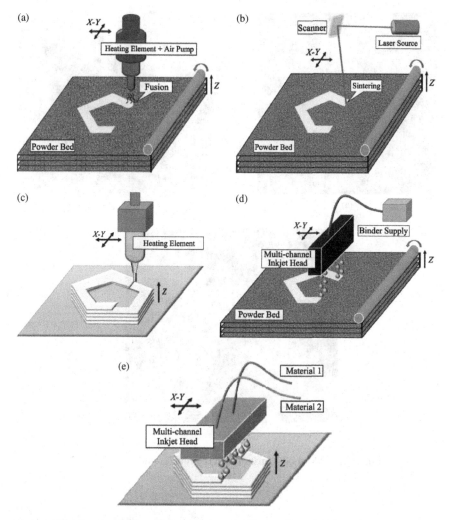

Figure 5.8 Food printers: (a) hot air-based, (b) laser-based, (c) melt-extrusion, (d) binder-jet, and (e) inkjet-based printers [19].

- **Inkjet printing:** Similar to binder-jet printing, the multi-material streams/droplets from a syringe-type apparatus are dropped on demand to create 3D edible food products, such as cookies, cakes, or pastries. It uses low viscosity paste or puree.

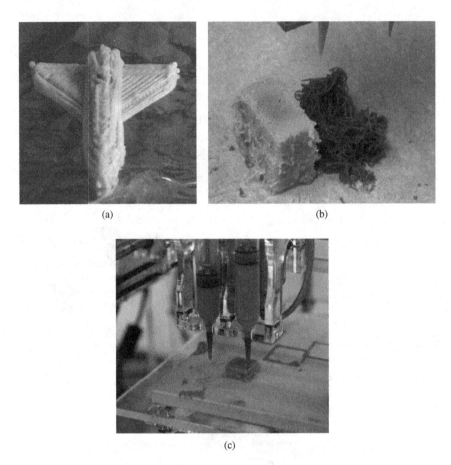

(a)   (b)

(c)

Figure 5.9   AM food items: (a) and (b) cookies and (c) turkey [22].

Some as-produced edible items are depicted in Figure 5.9. The printing precision and accuracy are critical to the application of AM technology in the food sector. One of the advantages is to fabricate an exquisite and fascinating structure of edible products to increase the consumer's interest and appetite. Therefore, to achieve precision and accuracy, material properties (rheological properties and particle size), process parameters (nozzle diameter, printing speed, and printing distance), and post-processing methods (baking, frying, and cooking) should be well optimized so that the taste of the products is not compromised [20]. As of now, the various commercial AM models for food printing are listed below [17]:

- Choc Creator uses FDM technology and creates 3D edible chocolate models. Its price is about £2,380.
- Power WASP EVO is also dedicated to chocolate printing. It costs less than $1,000.
- Zmorph cake and chocolate extruder is a very simple and cheap solution. It costs less than $350.
- XYZPrinting food printer is projected at around $2,000.
- ChefJet series printer costs about $1,000.
- Monochrome food printer costs about $5,000.
- CandyFab printer is used for sugar printing and is also cost-effective.
- Fab@Home printer is a multi-head printer and can print using a wide range of materials including grounded meats.
- Bocusini food printing system is an economical printer for sugar, chocolate, sweet jellies, pastries and marzipan, cheese, mashed potatoes and vegetables, as well as grounded meat. The cost of the printer is less than €1,200.
- Foodini printer can make meat, pasta, chocolate, cake, or mixed fruit. It costs around $1500.

This differentiated food cooking process provides an engineering solution to digitalized food design and nutrition control. However, the end products should be tested on various aspects to solve the technical bottlenecks. The current scenario of food-based AM technologies is developing incredibly fast and it would be wrong to judge the potential of these technologies on the basis of the feedstock food materials. As new feedstock and processing materials are being developed, it is believed that the technology will definitely enter real business in restaurants and food outlets [21].

5.2.4. *Wire-Arc AM*

The wire-arc AM (WAAM) technique has attracted the large-scale manufacturing industry as it is capable of both manufacturing products of metals and alloys as well as repairing fractured surfaces. The development of WAAM is being driven by the need for increased manufacturing efficiency of engineering structures. This setup is a result of implementing the "additive" principle in the wire-arc welding process. Since welding itself is a manufacturing operation wherein the heat energy is used to fuse the molten metallic pools one upon another, it is easy to upgrade by including the AM principle.

WAAM possesses high deposition rate of the metallic feedstock in the form of a wire filament, similar to conventional welding process. Basically, WAAM uses an electric arc as the primary source of heat. It is able to produce very near net shape preforms without the need for complex tooling, moulds or dies which saves significant cost and lead time reductions. Along with this, it enjoys increased material efficiency, improved component performance and reduction of inventory and logistics costs for on-demand manufacture. The very first patent on WAAM was granted in the 1920s; therefore it is the oldest and least recognized category of AM processes. Nevertheless, the system has been widely used for local repairs on damaged or worn components, and to manufacture round components and pressure vessels for several decades. In the recent past, with the advent of the highly efficient CAD and computer-aided manufacturing (CAM) software, WAAM has undergone significant developments. The timeline of the WAAM process is given below:

- In 1983, shape welding was used to manufacture high quality large nuclear steel parts.
- In 1993, Prinz and Weiss patented the *Shape Deposition Manufacturing* process that was included with computer numerical control (CNC) milling.
- At the end of the 20th century, Cranfield University patented the *Shaped Metal Deposition* process for developing engine casings using different materials.

As per the current market, following are the major players of the WAAM technique:

- Cranfield University in collaboration with various industries (WAAMAT programme)

- Norshk Titanium
- Gefertec Gmbh

Today's WAAM systems offer higher resolution of about 1 mm along with deposition rates ranging 1–10 kg/h. Furthermore, it offers numerous other benefits such as less material wastage, increased design flexibility, and reduction in lead time and manufacturing cost. A standard WAAM system consists of the following key components (refer to Figure 5.10):

- A three or five axes CNC gantry table
- Welding torch
- Filament wire
- Argon gas cylinders

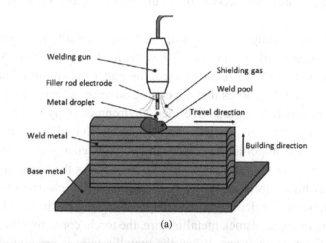

(a)

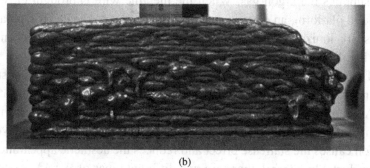

(b)

Figure 5.10 (a) Schematic of the WAAM process [23] and (b) as-produced component [24].

- Gas shielding
- Electric source
- Computer Interface

The process starts with the CAM software that is used to generate the tool path for the gantry table, command the start/stop points of welding, and decide the feed rate of metallic wire filament. The software is also capable of converting the CAD model to a printable code by following the slicing principle. Thereafter, the welding torch moves in the direction according to the coordinates and triggers the wire feeder to deposit the material in the path. Once the arc is ignited, the wire filament transforms into the molten weld pool in the given direction. The molten wire is extruded in the form of beads on the substrate and the beads stick together. The molten beads are deposited on the platform and allowed to cool down.

Then, the welding tool moves upwards equivalent to the layer thickness of the deposited melt pool to follow the next layer printing. After the required numbers of layers are printed, the part is taken out of the build chamber. It should be noted that depending on the reactivity of the molten metal characteristics, a protective inert environment may be provided to eliminate any type of possible contamination caused by atmospheric gases. Now, it is important to understand how the energy is actually produced for melting the feedstock wire. WAAM obeys the common principles of gas metal arc welding processes, wherein an electric polarity is first set between the deposition nozzle and the build platform. Alongside the feeding of the feedstock metallic wire, the torch, consisting the nozzle, also supplies the argon gas. When the metallic wire is brought closer to the build platform, an electric spark originates due to the electric polarity. The spark ionizes the gap between filament wire and build platform by utilizing the supply gas.

The ionized steam turns into plasma and carries the molten feedstock towards the build platform. It should be noted that post-processing of the WAAM product is always essential as the manufactured products need significant cleaning and machining operations. In addition, the layered fashion causes the staircase effect and reduces the aesthetic appearance of the products. In general, there are two different types of WAAM systems,

Figure 5.11   Industrial scale robotic WAAM system (adopted from TWI).

namely robotic or machine tool-based. However, any three-axis robotic arm and an arc welding power source can be combined to make an entry level WAAM system (refer to Figure 5.11 for an industrial scale robotic WAAM setup). Titanium, aluminium alloys, steels, invar, Inconel, and copper alloys are the most common class of feedstock metal filaments for the WAAM process. Landing gear ribs, wing spar, impellers, turbine blades, shrouds and linkages for various industries are the primary products. The various benefits, limitations, and applications of WAAM are discussed below:

- Benefits of WAAM
  - Cheaper process and materials
  - High mechanical strength of the parts
  - High deposition rate
  - Large scale manufacturing

- ○ Suitable for repair operations
- ○ Manufacturing flexibility
- ○ Wide range of metallic materials
- ○ Higher manufacturing speed
- Limitations of WAAM
  - ○ Residual stresses and distortions
  - ○ Poor surface finish
  - ○ Post-processing is required
  - ○ Poor dimensional accuracy
  - ○ Materials require shielding
  - ○ Low resolution

5.2.5. *4D Printing*

Four-dimensional (4D) printing facilitates AM objects to change their shape over time and to execute a purpose-based end-user phenomenon. 4D printing is also known as programmable technology that employs AM systems to manufacture smart feedstock materials to exhibit additional characteristics enabling it to change shape and respond to natural stimuli, such as temperature, moisture, etc. This emerging technology combines the different types of AM technologies with high-level material science, engineering, and design modifications. Indeed, the materials play an utmost important role in 4D printing, as the specially designed feedstock materials have to behave in accordance to the specific stimulus, in the form of heat, moisture, light, magnetic and electric fields, etc. For the first time in 2014, the 4D printing technology innovation grant was approved by MIT, USA to Tibbits' Self-Assembly Printing Lab. Since then, the lab, in collaboration with Autodesk, has worked on developing computer systems to allow geometrical inputs for measuring the changes incurred after external stimulus exposures [25].

Indeed, 4D printing is the very recent industrial revolution of AM to manufacture smart products. Hence, knowledge of the 4D printing concept is still growing as there is a lot of research going on to demonstrate potential breakthroughs of this wonderful technology. It is an interdisciplinary branch of AM technology that involves different aspects of materials, technology, and mathematical modelling. The as-produced 4D products can be used for a wide range of architectural, textile, food, defence, and transportation applications; however, bio-printed products have evolved the most beneficial application as it can save lives by printing organs of the human body. Since human organs respond to the body's environment, without any outside intervention, manufacturing biomedical products mimicking body environment can be achieved through 4D printing [26]. Nonetheless, 4D printing is still immature and suffers from the following challenges:

- As deformations applied to 4D printed products are frequent, the geometries can degrade with time.
- There are still nascent proof-of-concepts.
- Stimuli from the environment, human body, and other constraints are fluctuating, so only a dynamic system can respond with ease and agility.
- The CAD software may not be able to program and create materials with multifunctioning properties and adaptability to different environmental situations.

Several AM technologies can be used for the suitable processing of smart materials, such as SLA, FDM, binder jetting, ink-jet printing, and selective laser melting. By using these technologies, majority of the smart materials, as given below, can be easily processed:

- **Shape-memory polymers:** Shape-memory polymers are the materials that are capable of storing a macroscopic shape, preserve it, and again return to the original shape under the effect of external stimulus. Because of the actuation capability of shape-memory polymers, these can be used for a wide range of applications in aerospace, soft robotics, biomedical, and other fields.
- **Shape-memory alloys:** This is the class of smart metallic materials which exhibits shape memory and recovery characteristics. They are useful for aerospace, civil engineering, and biomedical devices.
- **Liquid crystal elastomers:** These are heat-sensitive liquid crystals which can expand and transform according to the dictated code of a heat source.

- **Hydrogels:** These are polymeric chains which undergo photopolymerization processes. Hydrogels are hydrophilic networks that can retain a large amount of water. These polymers are stimulus sensitive and can be used for a number of biomedical applications.

- **Ceramics:** The novel ceramic inks can be combined with shape-memory polymers where the ceramic precursors act as soft-links which can be stretched three times beyond their initial length. The promising applications of ceramics include electronic aerospace devices.

- **Multi-materials:** The feedstock systems for 4D printing can be composites of shape-memory polymers, hydrogels, carbon, and/or even wood fibers. The prime motive of multi-material composites is to enhance the mechanical properties and response time of the as-produced structures.

Figure 5.12(a) shows the shape change response mechanisms of various materials for 4D printing, while Figure 5.12(b) shows the pictorial view of shape-changing objects. 4D printed devices are the potential candidates for applications in unusual environments due to their vast customizability and absence of mechanical elements. Surgical treatments involving 4D printing have already been performed, successfully demonstrating its influence. Further advancements in printable smart materials, mathematical models, and printing technologies are believed to allow this technology in futuristic targeted drug delivery, soft robotics, and other unthought-of fields in engineering [27]. Polymer-based AM technologies, such as fused deposition modelling, stereolithography, inkjet, and binder-jet printing, are mainly suitable for constructing nano-scale architectures with conductive materials which have exhibited great potential for future perspectives of electronics, sensors, and other conductive devices. Particularly, metal-nanoparticle inks are used for minimizing the size of printed features. Further, the high efficacy of liquid metals and reactive inks further adds on the features to flexible electronic devices. On the other hand, metal-based AM processes are suitable for fabricating electrodes and conductive interconnects [28]. Often, the potential of 4D printing technology is always underestimated; other than the shape-memory effect of the finally produced devices, it exhibits other intrinsic

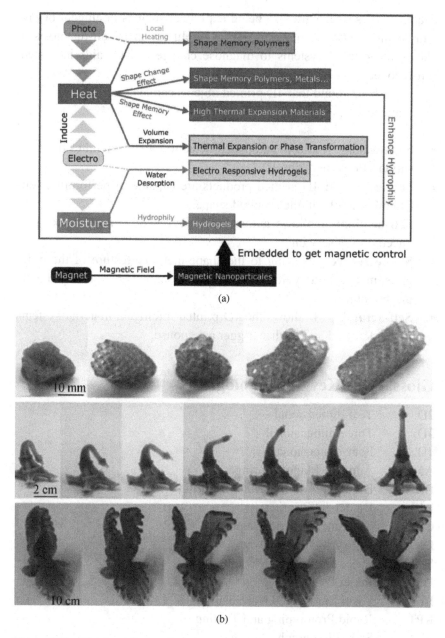

(a)

(b)

Figure 5.12 (a) Materials that can respond to heat, moisture, light, electricity, and magnetic fields [27] and (b) shape recovery of 3D objects (left to right) [29].

characteristics which are also being exploited, such as intelligent behaviour of the feedstock materials. An exemplification is self-diagnosis: it allows the printed systems to diagnose changes which aids in system self-recovery.

It has been found that 4D printed objects can also exhibit the following characteristics [30]:

- **Self-sensing:** It allows automated detection and sometimes quantification of external stimuli.
- **Self-assembly:** 4D printed products are capable of performing self-assembly to obtain the required shape.
- **Self-healing:** It allows automated recovery of minute cracks produced during servicing.
- **Shape recovery:** Owing to the shape-memory feature of the polymers, metals, and hydrogels, the original shape is retained by the 4D printed objects.
- **Self-actuating:** It allows the 4D printed intelligent material systems to get actuated and further trigger a response.

Glossary of Key Terminologies

2D: Two-dimensional
3D: Three-dimensional
4D: Four-dimensional
AM: Additive Manufacturing
BAAM: Big-Area Additive Manufacturing
BJP: Binder Jet Printing
CAD: Computer-Aided Design
DNA: Deoxyribonucleic Acid
FDM: Fused Deposition Modelling
IJP: Inkjet Printing
LENS: Laser Engineered Net Shaping
RPT: Rapid Prototyping and Tooling
SLA: Stereolithography
SLS: Selective Laser Sintering

.*STL*:   Standard Tessellation Language
UV: Ultraviolet
Z-Corp: Z Corporation

Exercise Questions

Q1: What is special purpose additive manufacturing? Write a short note on it.

Q2: Discuss bioprinting and different types of AM systems available for bioprinting.

Q3: Write any five potential applications of bioprinting.

Q4: What is food printing? How can edible items be cooked through AM?

Q5: What are the different types of food printing technologies available?

Q6: Define BAAM and discuss its benefits.

Q7: How can AM technology be used for construction applications?

Q8: What is the main difference between AM and 4D printing?

Q9: Discuss the key material category of 4D printing.

Q10: What are the key requirements of feedstock material for purpose-based AM?

References

1. Vyas D, Udyawar D. A review on current state of art of bioprinting. In 3D Printing and Additive Manufacturing Technologies. 2019 (pp. 195–201). Springer, Singapore.

2. Ozbolat IT, Yu Y. Bioprinting toward organ fabrication: Challenges and future trends. *IEEE Transactions on Biomedical Engineering*. 2013 Jan 30;60(3):691–699.

3. https://medicalfuturist.com/3d-bioprinting-overview/

4. Murphy SV, Atala A. 3D bioprinting of tissues and organs. *Nature Biotechnology*. 2014 Aug;32(8):773.

5. Lee VK, Dai G. Three-dimensional bioprinting and tissue fabrication: Prospects for drug discovery and regenerative medicine. *Advanced Health Care Technologies*. 2015;1:23.

6. Mandrycky C, Wang Z, Kim K, Kim DH. 3D bioprinting for engineering complex tissues. *Biotechnology Advances*. 2016 Jul 1;34(4):422–434.

7. Dababneh AB, Ozbolat IT. Bioprinting technology: A current state-of-the-art review. *Journal of Manufacturing Science and Engineering.* 2014 Dec 1;136(6).

8. Sudbury Z, Duty C, Kunc V, Kishore V, Ajinjeru C, Failla J, Lindahl J. Characterizing material transition for functionally graded material using big area additive manufacturing. *In Annu Int. Solid Free. Fabr. Symp.* 2016 Jan 1.

9. Post BK, Richardson B, Lind R, Love LJ, Lloyd P, Kunc V, Rhyne BJ, Roschli A, Hannan J, Nolet S, Veloso K. Big area additive manufacturing application in wind turbine molds. *In Solid Freeform Fabrication Symposium.* 2017 Aug 9.

10. Tekinalp HL, Kunc V, Velez-Garcia GM, Duty CE, Love LJ, Naskar AK, Blue CA, Ozcan S. Highly oriented carbon fiber–polymer composites via additive manufacturing. *Composites Science and Technology.* 2014 Dec 10;105:144–150.

11. Kishore V, Ajinjeru C, Nycz A, Post B, Lindahl J, Kunc V, Duty C. Infrared preheating to improve interlayer strength of big area additive manufacturing (BAAM) components. *Additive Manufacturing.* 2017 Mar 1;14:7–12.

12. Li L, Tirado A, Nlebedim IC, Rios O, Post B, Kunc V, Lowden RR, Lara-Curzio E, Fredette R, Ormerod J, Lograsso TA. Big area additive manufacturing of high performance bonded NdFeB magnets. *Scientific Reports.* 2016 Oct 31;6:36212.

13. Bos F, Wolfs R, Ahmed Z, Salet T. Additive manufacturing of concrete in construction: Potentials and challenges of 3D concrete printing. *Virtual and Physical Prototyping.* 2016 Jul 2;11(3):209–225.

14. Meibodi MA, Bernhard M, Jipa A, Dillenburger B. The Smart Takes from the Strong: 3d Printing Stay-In-Place Formwork for Concrete Slab Construction. Fabricate: Rethinking Design and Construction, UCL Press, London. 2017:210–217.

15. Craveiroa F, Duartec JP, Bartoloa H, Bartolod PJ. Additive manufacturing as an enabling technology for digital construction: A perspective on Construction 4.0. sustainable development. 2019 Jul 1;4:6.

16. Ma G, Wang L, Ju Y. State-of-the-art of 3D printing technology of cementitious material — an emerging technique for construction. *Science China Technological Sciences.* 2018 Apr 1;61(4):475–495.

17. Izdebska J, Zolek-Tryznowska Z. 3D food printing–facts and future. *Agro FOOD Industry Hi Tech.* 2016 Mar 1;27(2):33–37.

18. Millen C, Gupta GS, Archer R. Investigations into colour distribution for voxel deposition in 3D food formation. In 2012 International Conference on

Control, Automation and Information Sciences (ICCAIS). 2012 Nov 26 (pp. 202–207). IEEE.

19. Sun J, Peng Z, Yan L, Fuh JY, Hong GS. 3D food printing an innovative way of mass customization in food fabrication. *International Journal of Bioprinting*. 2015 Jul 2;1(1).

20. Liu Z, Zhang M, Bhandari B, Wang Y. 3D printing: Printing precision and application in food sector. Trends in Food Science & Technology. 2017 Nov 1;69:83–94.

21. Pallottino F, Hakola L, Costa C, Antonucci F, Figorilli S, Seisto A, Menesatti P. Printing on food or food printing: A review. *Food and Bioprocess Technology*. 2016 May 1;9(5):725–733.

22. Lipton J, Arnold D, Nigl F, Lopez N, Cohen DL, Norén N, Lipson H. Multi-material food printing with complex internal structure suitable for conventional post-processing. *In Solid Freeform Fabrication Symposium*. 2010 Aug 9:809–815.

23. Ding D, Pan Z, Van Duin S, Li H, Shen C. Fabricating superior NiAl bronze components through wire arc additive manufacturing. *Materials*. 2016 Aug;9(8):652.

24. Cunningham CR, Flynn JM, Shokrani A, Dhokia V, Newman ST. Invited review article: Strategies and processes for high quality wire arc additive manufacturing. *Additive Manufacturing*. 2018 Aug 1;22:672–686.

25. Kwok TH, Wang CC, Deng D, Zhang Y, Chen Y. Four-dimensional printing for freeform surfaces: Design optimization of origami and kirigami structures. *Journal of Mechanical Design*. 2015 Nov 1;137(11).

26. Ramesh S, Usha C, Naulakha NK, Adithyakumar CR, Reddy ML. Advancements in the research of 4D printing-a review. In IOP Conference Series: Materials Science and Engineering. 2018 Jun (Vol. 376, No. 1, p. 012123). IOP Publishing.

27. Zhang Z, Demir KG, Gu GX. Developments in 4D-printing: A review on current smart materials, technologies, and applications. *International Journal of Smart and Nano Materials*. 2019 Jul 3;10(3):205–224.

28. Chang J, He J, Mao M, Zhou W, Lei Q, Li X, Li D, Chua CK, Zhao X. Advanced material strategies for next-generation additive manufacturing. *Materials*. 2018 Jan;11(1):166.

29. Zarek M, Layani M, Cooperstein I, Sachyani E, Cohn D, Magdassi S. 3D printing of shape memory polymers for flexible electronic devices. *Advanced Materials*. 2016 Jun;28(22):4449–4454.

30. Li X, Shang J, Wang Z. Intelligent materials: A review of applications in 4D printing. *Assembly Automation*. 2017 Apr 3.

Chapter 6

Testing and Measurement

The present chapter focuses on the testing and measurement of Additive Manufacturing (AM) components. Apart from this, it also outlines the procedures for performing the basic laboratory experimentation.

6.1. Importance of Testing and Measurement

Scientific and engineering societies across the world are continuously testing and measuring the already existing as well as newly manufactured components. The importance of the various types of testing and measurement approaches increases as every commercial organization is trying to advocate the superiority of their products. Therefore, there is a need for standardised scientific methods to produce measurable outputs, either in quantitative or qualitative terms. For instance, a simple observation would be the thickness of the modern era smartphone devices or smart television. Indeed, this can be simply done by using a steel foot rule; however, the authenticity of the measured thickness will be questionable. This is because the steel foot rule may not be able to maintain the precision of the quantified thickness upon repeating the activity. Another issue will be the limitation in terms of minimum value of the measurement. As the steel foot rule has a minimum accuracy level of 1 mm, any dimension below 1 mm cannot be measured with it. This is important as there are numerous engineering and scientific products which have sizes ranging below 1 mm, for example thickness of a notebook paper, plastic film, etc. This makes testing and measurement one of the most fundamental concepts in engineering as it helps scientists to conduct experiments or form new theories. Moreover, testing and measurement are essentially important in agriculture, construction, commerce, and even in daily routine. The word "measurement" comes from the Greek word "metron," which means "limited proportion". Therefore, the word "measurement" means a scientifically matured technique by which the properties of an object are determined by comparing them to a standard. Indeed, measurements require tools and gadgets to provide scientists with a quantity. On the other hand, testing means to check the abilities of the manufactured goods, for instance highest speed range of a sport car. It includes standard protocols which are generally executed while obtaining the quantified data that can categorize the goods on the basis of their capabilities. Before moving ahead, it is important to understand the following terms:

○   *Object:* It is an existing and newly manufactured device that is under observation.
○   *Tool:* The device used for measuring the characteristics of the object.
○   *Measurement:* It is the activity of obtaining the quantitative or qualitative data about an object by using a tool.
○   *Standard:* For validation, the measured quantity of an object is compared with a database that is known as the standard.

Testing and measurement play a significant role in an industrial environment as their inaccuracy will lead to the commercialization of defective products in the market, which could have an adverse impact on the market value of an organization. Therefore, the basic principle of correct measurements is to ensure that all personnel involved in manufacturing are skilled and can correctly perform the testing and measurement. In addition, the various testing and measurement activities must involve standard protocols set by internationally acknowledged societies, for instance International Organization for Standardization (ISO) and American Society for Testing and Materials (ASTM), so that the qualitative and quantitative characteristics of the industrial goods can be repeatable irrespective of the geographical location. Measurement and testing are the two distinct activities as described below:

- **Measurement:** Measurement is defined as a branch of engineering that involves production of the measurable quantitative and qualitative results by adopting the principles of metrology. Metrology is the science of measurement, encompassing all theoretical and experimental aspects and can be categorized into scientific, applied, and legal metrology. The overwhelming portion of most metrology applications is carried out by engineers and technicians. It is often recommended that a metrology engineer should have adequate knowledge and skill sets in operating a wide range of engineering tools. However, there exist numerous factors which can reduce the accuracy or precision of measured results, such as:

  ○ *Environmental conditions:* Any change in the environmental conditions such as temperature and humidity tends to expand and contract materials as well as affect performance of the measurement equipment.

  ○ *Inferior measuring equipment:* Equipment or tools that have been poorly maintained, damaged, or not calibrated will give unreliable results. It follows the repeatability and reproducibility principles which states that under the same environmental conditions as well as similar levels of human expertise, the equipment must produce same and close results, respectively, for each consecutive measurement.

  ○ *Poor measuring techniques:* If the metrology engineer or technician doesn't follow consistent procedures for measurements, errors can arise.

  ○ *Inadequate staff training:* Inadequate staff training on how to use equipment and which principle to follow can result in vague outcomes.

There are different types of metrological measurements involved, for example measuring the geometrical dimensions (such as linear and angular), roundness, profile of the intricate shapes, surface roughness, and other topological parameters, etc. Common classes of tools used for measuring the aforementioned properties are described below:

- ○ *Vernier Calliper:* This is a metrological instrument that is basically used for measuring linear dimensions. The least count of manual and digital Vernier calliper is 0.02 and 0.01 mm, respectively.
- ○ *Depth Gauge:* This tool is used to measure the depth of any open blind or through hole. It obeys the Vernier Calliper principles and has a least count of 0.01 mm.
- ○ *Micro-meter:* The micro-meter is used to measure the geometrical dimensions of round products. The least count of a manual and digital micro-meter is 0.01 and 0.001 mm, respectively.
- ○ *Optical Profilometer:* This device is used to observe the profiles and shapes of tiny devices, such as the gears of wristwatches. It can provide the gear angle, diameter, and pitch.
- ○ *Sine-bar and Slip Gauges:* These are the two different devices which are generally used together for measuring the angle of any geometrical surface or machine tool by adopting the sine principle of trigonometry.
- ○ *Surface Roughness Tester:* This digital device is used to measure the smoothness or roughness of surfaces produced by different manufacturing process.

- **Testing:** Mechanical testing is an essential part of any design and manufacturing process. The scope of mechanical testing ranges from characterizing material properties to validating the final products. Safeguarding protection is the prime aim of the mechanical testing; however, it also plays an important role in producing a cost effective design and in its technological evolution. The fundamental procedure of performing any kind of test is generally established by the government agencies, such as ASTM and ISO. The required test methodologies are typically specified in terms of established drawings to follow for producing a physical component ready for actual testing protocol. Qualified design and manufacturing engineers understand the value of testing to validate models, improve production, and increase the overall quality of parts. The time and expense of testing might seem like an easy place to reduce costs, but surprising failures after parts are already in service are much more difficult to bear. Typical mechanical tests include, but are not limited to the following:

○ Tensile testing of a product to ensure performance
○ Hardness testing of final parts to observe the requirements of heat treatment processes
○ Fatigue testing to establish long term serviceability
○ Model validation to streamline the design and manufacturing processes
○ Fracture mechanics tests to support futuristic maintenance schedules
○ Combination of mechanical and model validations to understand the type of material required.

6.2. Types of Testing

This section of Chapter 6 describes the different types of non-destructive and destructive testing procedures available for validating the qualitative and quantitative features of manufactured goods.

• **Non-destructive Testing:** Non-destructive testing, often abbreviated as NDT, refers to the function of quality control and corresponds to recognized methods which have been developed over extended stretches of time. NDT can be defined as the testing of materials for their external and internal flaws, metallurgical conditions, and defects which cannot be seen by the naked eye, without interfering with the integrity of the material or its suitability for service. This means that the materials under observation are not undergoing any physical damages. The NDT approach is applied on a sampling basis by selecting a few items out of the whole stock manufactured in a particular interval of time. In this way, the whole produced lot size can be certified upon appraising outcomes. NDT is a high-end technology concept used to evaluate the products in a robust manner and can be applied to any intermediate stage of an industrial environment. Indeed, a certain degree of skill sets is normally required to apply the NDT techniques by adopting a standard protocol so that maximum amount of information concerning the product can be obtained for consequential feedback. NDT should not be considered as only a technological method of rejecting or selecting the produced goods on the basis of any mismatch or match, respectively. In actuality, it provides a lot more information about the reasons responsible for assurance about the suitability of various engineering facilities utilized in production. NDT is a very broad and interdisciplinary field that plays a critical role in ensuring that structural components and systems perform their function in a reliable fashion. Depending on the type of industry and product, different NDT approaches are followed to satisfy all requirements in all circumstances. Certain standards have been also implemented to assure the reliability of the NDT tests and prevent certain errors due to either the fault in the equipment used, the misapplication of the methods or the skill and knowledge of the inspectors. The different methods covered in NDT are discussed below [1]:

- *Visual Inspection:* The visual inspection of the surface and dimensions of any produced object gives qualitative output and is only limited to macroscopic errors. Therefore, this method will not be able to identify internal defects. Essentially, visual inspection should be performed in the way that a car salesman inspects a car out for delivery. This is helpful to spot dimensional distortions, rough surfaces, missing materials and operations, poor fits, large cracks, cavities, unevenness, wrong parts, etc.

- *Microscopic Inspection:* In this method, the external porous holes, cracks, and surface irregularities are observed with magnified optical lenses. This method gives qualitative outputs of the surface under observation. As compared to the previous one, it is a more precise approach as it also provides micro-images to conduct internal and external feedbacks.

- *Radiography:* This is the method generally used to identify cracks and internal defects which cannot be seen with the naked eye. This is a comparative test wherein a defective product has been compared with a sound product. In this test, X-rays emitted from a source penetrate the product surface as a function of the accelerating voltage. The existence of any internal void or cavity allows the passage of more X-rays that in turn let the film under the test object to be exposed to more rays (refer to Figure 6.1(a)). The exposed film will draw a pattern illustrating the geometry and size of the internal defect. This test

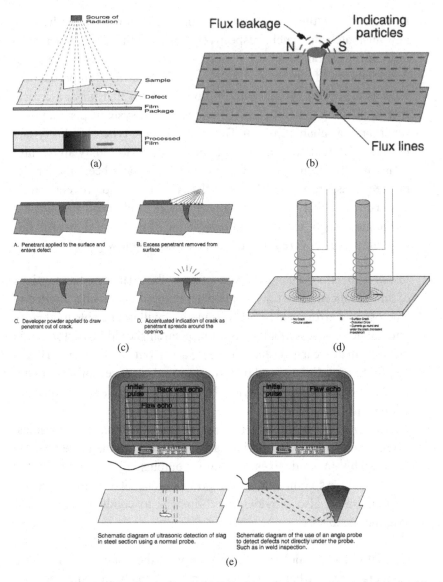

Figure 6.1 Schematic illustrations of NDT: (a) radiography, (b) magnetic particle inspection, (c) liquid die penetration, (d) eddy current, and (e) ultrasonic flaw detection tests [1].

can be done on thick products, ranging 20–25 mm, and is capable of identifying tiny defects of about 0.5 mm in dimension.

- *Magnetic Particle Inspection:* In this method, magnetic fields and small magnetic iron particles are used to detect the flaws in components. However, this method is suitable for only ferromagnetic materials, including iron, nickel, cobalt, and their alloys. During the test (refer to Figure 6.1(b)), an electromagnet yoke is placed on the object surface and a kerosene-iron filling suspension is poured on the surface after activating the electromagnet. If the surface under observation consists of any crack or flaw, the magnetic flux will be broken into new south and north poles. In turn, the iron particles will also be attracted at the edges of the crack. Thereby, the obtained cluster of iron particles can be seen easily and the defects can be identified. In order to obtain the best sensitivity, it is recommended to use the lines of magnetic force perpendicular to the defective surface.

- *Liquid Die Penetration Test:* This test is relatively easy and flexible. It utilizes capillary action to locate minute microscopic defects on the exterior of the manufactured products. Despite being inexpensive, the liquid die used in this method is unable to inspect subsurface flaws and generally loses resolution when applied to porous materials. The most suitable materials for this method are aluminium, copper, steel, titanium, glass and ceramic materials, rubber, and plastics. This method is used to inspect fatigue, quenching, grinding, overload, and impact cracks. The object under inspection is first chemically cleaned to remove all traces of foreign material and then bright red or ultraviolet fluorescent die is applied to penetrate the surface for about 15 min (refer to Figure 6.1(c)). After this, the excess of die is removed and a thin chalk powder coating is applied. The applied coating is allowed to rest onto the surface for the development time and then is taken away. The chalk coating also takes away the penetrated die from the cracks. The dried die outlines the magnified visual of the crack dimensions and geometry.

- *Eddy Current test:* Like the previous NDT technique, the eddy current test also detects surface or subsurface flaws. It is sensitive to the conductivity of the material, its permeability, and dimensions. Eddy current is generally induced in an electrically conducting material by subjecting it to an alternating magnetic field, ranging 10 Hz to 10 MHz. As the alternating current is applied to the conductor, an alternating magnetic field develops in and around the conductor. Now, when an another electrical conductor is brought closer to the changing magnetic field, current is also induced in the second conductor whose value is generally influenced by the presence of voids and cracks (refer to Figure 6.1(d)). The current forms impedance on the second coil which is used as a sensor. In the actual test, a probe is placed on the surface under inspection and electronic equipment monitors the eddy current in the work piece through the same probe. The alternating magnetic field is normally generated by passing an alternating current through a coil. Any change in the material or geometry can be detected by the excitation coil as a change in the coil impedance.

- *Ultrasonic Flaw Detection test:* This test uses a high frequency sound energy to inspect flaws, dimensional measurements, and material characterization. For contact testing, the oscillating crystal incorporated in a handheld probe is applied to the surface of the material to be tested (refer to Figure 6.1(e)). To facilitate the transfer of energy across the small air gap between the crystal and the test piece, a layer of liquid, usually oil, water, or grease, is applied to the surface. The handheld probe is capable of emitting and receiving the pulses in the form of mechanical oscillations as well as electrical pulses. In this way, the time taken by an electric pulse to pass through the test surface and its ability to return back will be interfered by the existing internal defects and flaws.

Table 6.1 shows the capabilities and limitations of the various NDT approaches.

It has been found that the visual inspection technique is generally applicable to preliminary part analysis and overall quality interpretation in AM applications. This technique is suitable for detecting surface discontinuities of manufactured parts that can also be employed during the printing process. Furthermore, the use of liquid die penetration test on AM parts can be performed without any requirement of polishing. Magnetic particle penetration test is one of the commonly applied NDT techniques for the casted parts but its application on AM is limited as it is only

Table 6.1 Capabilities and limitations of NDT approaches.

NDT Method	Applicable Standard(s)	Suitable Materials*	Capabilities	Limitations
Visual Inspection	—	P, M, and C	This method can identify macroscopic surface flaws in a cost-effective and rapid way	It cannot detect microscopic and subsurface flaws
Microscopic Inspection	ASTM E407	P, M, and C	Microscopic surface flaws can be easily observed using this approach	It is not able to detect subsurface flaws
Radiography	ASTM E1032-19	P, M, and C	This approach can detect subsurface flaws	The smallest defect that can be observed should be 2% of the product's thickness
Magnetic Particle Penetration	ASTM E3024/ E3024M-19 and ASTM E709-15	M_c	This method can identify surface, near surface, and layered flaws	It has limited ability to identify subsurface flaws and is only applicable to ferromagnetic materials
Liquid Die Penetration	ASTM E165/ E165M–18	P, M, and C	This technique is applicable to identify surface flaws only	It cannot be used to identify subsurface flaws and is unsuitable for porous materials
Eddy Current Test	ASTM E2884-17		Surface and near surface flaws can be identified with high sensitivity	It is only applicable to metallic materials
Ultrasonic Flaw Detection Test	ASTM E114-15	M_c	This technique is mainly used for identifying subsurface flaws with high accuracy	The test materials must be good conductors of sound

*Note: P, M, C, and M_c refers to polymers, metals, concrete, and conductive metals.

applicable for ferromagnetic materials. The use of eddy current method and ultrasonic flaw detection method are also limited for AM [2].

- **Destructive testing:** Destructive testing (DT) is generally performed to understand the performance characteristics of manufactured products in order to certify whether the as-manufactured product is suitable for a desired end-user application. Similarly, it also helps in understanding the effect of various design iterations, material combinations, and pre/ post-treatment processes on the specimen's performance. The word 'destructive' implies the undergone products are permanently damaged through the application of external forces or harsh environment. DT procedures can either follow specific standards or can be tailored to reproduce set service conditions. These methods are commonly used for materials characterisation, fabrication validation, failure investigation, and can form a key part of engineering critical assessments. Followings are the most essential DT approaches for the mechanical characterization of engineering designs, products, and materials:

- *Tensile Test:* As per the American Society of Metals (USA), the tensile test is generally preferred for ascertaining specifications of the material for quality services. The tensile test is generally recommended while developing new materials and processes so that a competitive study of the different types of materials and processes can be performed. In this test, the behaviour of a material under a uniaxial tensile force is studied. Further, the test provides a wide range of outputs, such as elastic point, yield point, and fracture point. In quantitative terms, it gives the load sustained and elongation attained by the material during testing. Using this data, a comparative plot between stress and strain could be obtained. Therefore, the test is also useful for ductile materials which can be elongated to a high extent, such as gold, copper, and silver. Figure 6.2(a) shows the schematic representation of a tensile test specimen, while Figure 6.2(b) gives the illustration of a tensile testing machine [3]. The workpiece is clamped in the jaws of the machine before a gradual load is applied to deform the workpiece. An example of stress vs. strain plot attained is given in Figure 6.2(c). The stress and strain of the workpiece are then calculated by using Equations 1 and 2, respectively:

$$S = P/A_o \qquad (1)$$
$$E = (L_f - L_o)/L_o \qquad (2)$$

where S, P, A_o, E, L_o, and L_f are the stress, load, original cross-sectional area, strain, original length of the specimen, and final length of the specimen, respectively.

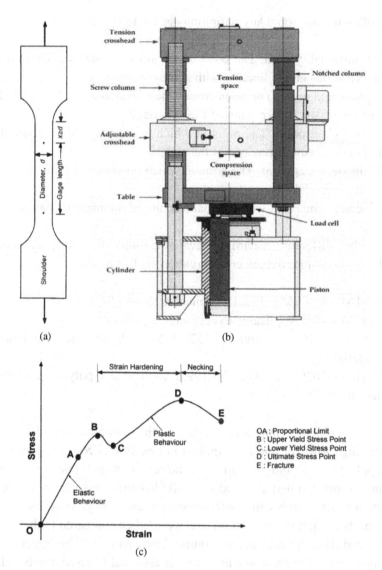

(a)

(b)

(c)

OA : Proportional Limit
B : Upper Yield Stress Point
C : Lower Yield Stress Point
D : Ultimate Stress Point
E : Fracture

Figure 6.2 (a) Design of a tensile test sample, (b) tensile testing machine [3], and (c) stress vs. strain plot.

Following are some key definitions of the tensile test:

✓ Proportionality limit: The point on the stress vs. strain diagram up to which stress grows linearly with strain
✓ Upper yield point: The point from which maximum load is applied to initiate plastic deformation of the material
✓ Lower yield point: The point at which minimum load is required to maintain plastic behaviour of the material
✓ Ultimate stress point: The point at which maximum load is sustained by the material
✓ Fracture: The point where actual fracture of the material happens

Further, different standards are available for tensile testing, depending on the types of material under observation, as follows:

✓ ASTM E8 / E8M – 16ae1: Applicable to metallic materials
✓ ASTM C496: Applicable to concrete
✓ ASTM D638 – 14 and ISO 527-1:2012: Applicable to polymeric materials
✓ ASTM D3039 and ISO 527-4:1997: Applicable to polymeric composite materials

Compression Test: Compression testing of the materials is similar to tensile testing; however, the load applied in this case is compressive. Upon the application of compression load, the relationship between stress and strain is similar to that obtained in tensile loading. This means that up to a certain value of stress, the material behaves elastically. After this, plastic flow starts as high value of strain develops. When compared to the tensile test, it is difficult to conduct the compression test due to the larger cross-sectional area of the test specimen that is essential to resist any buckling due to bending. The workpiece undergoes strain hardening with deformation and ultimately starts to take on higher load values.

It is of utmost importance to avoid the buckling action by selecting the ratio of height h to diameter d of the specimen to be less than 2. Figure 6.3 shows the schematic of a compression test specimen. The stress and strain of the workpiece are then calculated by using Equations 3 and 4, respectively:

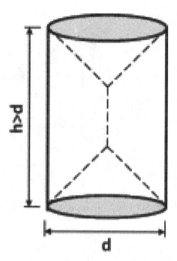

Figure 6.3 Schematic of compression test specimen [4].

$$S = P/A_o \tag{3}$$
$$E = (D_f - D_o)/D_o \tag{4}$$

where D_o and D_f are original and final diameter of the specimen respectively.

Further, different standards are available for compression testing, depending on the types of material under observation, as follows:

✓ ASTM E9 – 19: Applicable to metallic materials
✓ ASTM C39 / C39M – 20: Applicable to concretes
✓ ASTM D695 – 10 and ISO 604: Applicable to polymeric materials
✓ STM D3410 / D3410M – 16 and ISO 14126: Applicable to polymeric composite materials

• *Impact Test:* Another type of DT is where the impact strength of a notched specimen is tested under a sudden load applied by a swinging pendulum hammer to impart a fracture on the workpiece. The energy stored by the specimen material is then recorded to determine the fracture strength. Indeed, brittle materials will possess limited toughness due to their lower plastic deformation capability. On the other

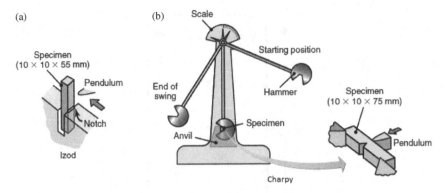

Figure 6.4 Schematic of impact test: (a) Izod and (b) Charpy (adopted from Green Mechanic).

hand, ductile materials have higher toughness as well as higher plastic deformation capability. The impact toughness test simulates the service conditions of component systems used in transportation, agriculture, and construction. A high impact resistance material is said to be a tough material. As the impact toughness is the capability of a material to sustain both fracture and deformation, a combination of strength and ductility are therefore essential. There are two different standardized tests, namely Charpy and Izod, which have been designed and used extensively to measure the impact energy. The apparatus for performing impact tests is illustrated schematically in Figure 6.4. As can be seen, the load is applied due to the impact blow of the weighted pendulum hammer, released from a fixed height h. The specimen is fixed at the base and knife edge of the pendulum strikes and fractures the specimen at the notch. After fracturing the test material, the pendulum continues to swing and rise to a height h', lower than the initial height of the pendulum. Therefore, the energy stored by the pendulum is obtained by using Equation 5:

$$E = mg(h - h') \tag{5}$$

where E, m, and g are energy stored by the work material, mass of the pendulum, and gravitational acceleration, respectively.

Further, different standards are available for impact testing, depending on the types of material under observation, as follows:

✓ ASTM E23 – 18 and ASTM E2248 – 18: Applicable to metallic materials
✓ ASTM C1747 / C1747M – 13: Applicable to concretes
✓ ASTM D6110 – 18: Applicable to polymeric materials
✓ ASTM D7136 / D7136M – 15: Applicable to polymeric composite materials

• *Fatigue Test:* Fatigue failure is tested to certify the behaviour of engineering materials when subjected to cyclic loads. Fatigue failures are the results of the repeated applications of stress, often below static yield stress. However, the reserved stress developed in the material tends to increase with each cycle and causes aging, as a result of stress concentration. As per ASM International [5], fatigue consists of three stages: (i) initial fatigue damage leading to crack nucleation and crack initiation, (ii) progressive cyclic growth of a crack, and (iii) sudden fracture of the cross section. There are three different categories of cyclic loadings, namely tensile-tensile, compressive-compressive, tensile-compressive. It would be possible to vary multiple which can influence the fatigue life and to state their effects through simplifying and idealizing the test conditions. However, there are still some unknown and uncontrollable factors which produce data scattering during a fatigue life test. There are generally two different classifications of fatigue life testing:
✓ Constant-amplitude test: This test module involves a sequence of amplitude obtained by applying reversals of stress of constant amplitude to the test-piece until it fails; refer to Figure 6.5(a), where R is the stress ratio.
✓ Variable-amplitude test: In this test module, variable sequences of amplitude are required in order to simulate the stresses to which a specimen is subjected in actual service conditions; refer to Figure 6.5(b).

Similar to other DT, different ASTM standards available are given below for the fatigue test:

✓ ASTM E466 – 15: Applicable to metallic materials
✓ ASTM C1361 – 10(2019): Applicable to concretes

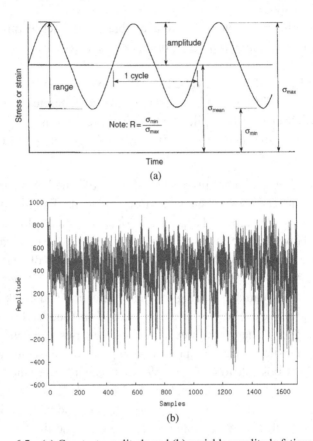

Figure 6.5 (a) Constant amplitude and (b) variable amplitude fatigue test.

✓ ASTM D7791 – 17: Applicable to polymeric materials
✓ ASTM D3479 / D3479M – 19: Applicable to polymeric composite materials

- *Flexural Test:* The flexural test, also known as three-point bending test, is an established standard protocol to test the flexural or bending strength of the test material. The test material of required dimensions is supported at the two ends and then a notch-shaped hammer pushes the test material to bend until the failure point. Figure 6.6 shows the schematic description of a three-point bending test. During testing, two different types of stresses are developed: (i) tensile stress on the bottom surface and (ii) compressive stress on the top surface [6]. It is

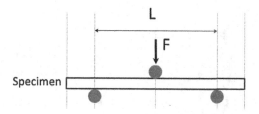

Figure 6.6   Schematic of the three-point bending test.

important to understand that there is a neutral axis, halfway from either surface. In general, most of the materials fail due to tensile stress prior to their failure due to compressive stress. Equation 6 is generally used for calculating the flexural strength:

$$\text{Flexural strength } (F_s) = P \times L/bd^2 \tag{6}$$

where P, L, d, and b are the load, span length, depth, and breadth of the workpiece, respectively.

Followings are the ASTM standards for different types of engineering materials:

✓ ASTM E855 – 08(2013) and ASTM E290: Applicable to metallic materials
✓ ASTM C78 / C78M – 18: Applicable to concretes
✓ ASTM D790 – 17: Applicable to polymeric materials
✓ ASTM D7264 / D7264M – 15 and ISO 899-2:2003: Applicable to polymeric composite materials

• *Hardness Test:* The hardness of a material is the property by which it resists the formation of a scratch under static or dynamic load. In material science, hardness is essentially one of the most important properties for a wide range of engineering applications. As the material is subjected to a static or dynamic load, it generally tends to oppose damages caused by the load application. As a consequence, harder materials can perform better as compared to ductile materials. There is a wide range of hardness testing protocols; however,

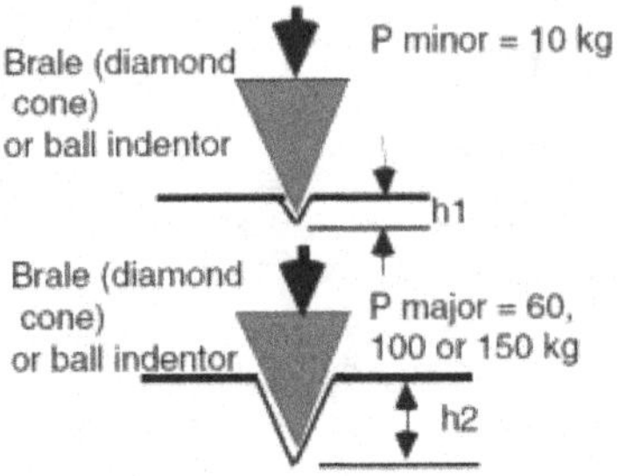

	Rockwell Scale (X =)	Indentor	Pmajor (kg)
	A	Brale (diamond)	60
	B	1/16" ball	100
	C	Brale (diamond)	150
	D	Brale (diamond)	100
	E	1/8" ball	100
	F	1/8" ball	60
	M	1/4" ball	100
	R	1/2" ball	60

Figure 6.7   Schematic of the Rockwell hardness test [8].

Rockwell and Brinell hardness procedures provide bulk hardness characteristics. As per ASM International, there are no absolute standards of hardness. Furthermore, it is not a quantitative property, except in terms of a given load applied in a specified manner for a specified duration and a specified penetrator shape is used [7].

Rockwell hardness: It is the test method that consists of indenting the test material with a diamond-shaped cone or hardened steel ball. The indenter is forced into the material by applying a preliminary load F_0 of about 10 kg (refer to Figure 6.7(a)). The indenter keeps on penetrating and finally reaches an equilibrium point, a datum position. After this, the preliminary minor load remains activated, while a major load is applied to increase the depth of penetration (refer to Figure 6.7(b)). As the equilibrium point is reached again, the applied major load is removed; however, the preliminary minor load still remains active. The removal of the major load allows partial recovery of the deformation. The permanent increase in depth of penetration is then calculated as the Rockwell hardness number, by using the following equation:

$$R_x = M - ((H_2 - H_1)/0.002) \tag{7}$$

where M is 100 for A, C, & D and 130 for B, E, F, M, & R; H_2 and H_1 are the final depth of penetration attained by the application of the additional load and datum point depth attained by the application of minor load, respectively.

Following are the Rockwell hardness ASTM standards for different types of engineering materials:

✓ ASTM E18 – 20: Applicable to metallic materials
✓ ASTM D785 – 08(2015): Applicable to polymeric materials
✓ ASTM F1957 – 99(2017): Applicable to polymeric composite materials

Brinell hardness: In this case, the Brinell hardness number (BHN) of hard, moderately hard, and soft material is determined. However, this method is not suitable to test very hard and brittle materials. Figure 6.8 shows the schematic description of the Brinell hardness test. The BHN is obtained by calculating the ratio of load applied to the spherical area of the impression. As the deformation caused by the indentation is similar to the tensile testing characteristics, by using empirical relationships, equivalent ultimate tensile strength of metals and alloys can be calculated [8]. Equations 8 and 9 are used for calculating BHN and ultimate tensile strength of the materials:

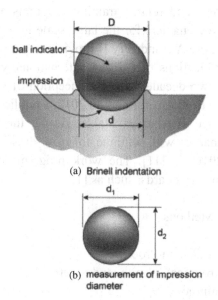

(a) Brinell indentation

(b) measurement of impression diameter

Figure 6.8 Schematic of the Brinell hardness test [8].

$$\text{BHN} = P/\left(\frac{\pi D\left(D - \sqrt{D^2 - d^2}\right)}{2}\right) \tag{8}$$

$$\text{Ultimate tensile strength} = 3.45 \times \text{BHN} \tag{9}$$

ASTM E10 – 18 is the only standard applicable to metallic materials.

6.3. ASTM/ISO Standards for Additive Manufacturing

As per the American Standards National Institute [9], AM technology incorporates a wide range of processes wherein a physical product is produced by using a digital computer-aided design (CAD) model. The complete process of producing AM parts includes consecutive steps: (i) designing the CAD model; (ii) specifying materials; (iii) establishing build parameters; (iv) controlling the AM build process; (v) post-processing; (vi) testing; (vii) certifying the part's fitness; and (viii) maintaining and repairing machines, parts, and systems. Indeed, the existence of standards, specifications, and related training programs are integral to AM process and are key enablers for the large-scale introduction and growth of AM. However, the AM industry is provisional on products, processes, and material certifications that follow internationally recognized standards [10]. About two decades after the inception of AM technology, the ASTM committee F-42 was formed. In 2012, a public-private partnership came into existence as America Makes. As per the Wohlers Report in 2018, the AM market will continue to grow drastically with an 80% increase from 2016–17 [11]. The work program of F42 has already approved numerous standards, such as [12]:

- F42.01: Test Methods
- F42.04: Design
- F42.05: Materials and Processes
- F42.06: Environment, Health, and Safety
- F42.91: Terminology
- F42.95: U.S. TAG to ISO/TC 261

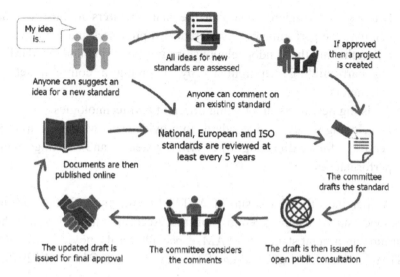

Figure 6.9 Standard development process (adopted from MTA Standards Update Booklet).

The development of a standard protocol is described in Figure 6.9. It is expected that new and existing ASTM standards on AM will be multi-branded. This is seen as a significant development because it should reduce, perhaps even eliminate, conflicting and competing international standards on AM. The ASTM technical committee identified as Additive Manufacturing Standardization Collaborative (AMSC) is continuously working on developing various standards to be applicable for AM. Further, the European Committee for Standardization (CEN) has also approved various standards, in collaboration with ASTM and ISO. Apart from this, there are various ventures, for instance AM's Center of Excellence, American Welding Society, Institute for Electrical and Electronics Engineers, Association Connecting Electronics Industries, Medical Imaging & Technology Alliance, Digital Imaging and Communications in Medicine, Metal Powder Industries Federation, MTConnect Institute, and SAE International for the development of various protocols applicable to the various engineering and scientific sectors. All such ventures are developing consensus standards to support the adoption of AM across multiple industry sectors by [13]:

- Establishing standards enabling the manufacturers to observe and compare the performance of different AM processes and materials
- Strengthening the vendor relationships by specifying and standardizing part-building requirements by obtaining a common set of parameters
- Assisting new users in adopting different AM technologies
- Developing widespread standards to provide users with uniform procedures for calibrating the AM systems and testing their performances

As per the official website of ASTM (www.astm.org), the various developed standards are given in Table 6.2. According to the website, the four most important standards of AM, especially for the novices, are ISO/ASTM 52900, ASTM 2924, ISO/ASTM 52901, and ISO/ASTM 52910.

- **ISO/ASTM 52900-15:** This standard defines the standard terminology of AM systems and is a successor to ASTM F2792 that was withdrawn. It is the first standard for AM that has been jointly developed and accepted by the ISO and ASTM, and subsequently accepted by the CEN. With this standard, the very first step was taken for uniting the world's standards bodies around AM. Another revision of this standard is currently being developed and will contain much more detailed guidelines and further clarifications for the industry. This standard includes the important terminologies, their descriptions, nomenclature, and different acronyms associated with the system and testing methodologies for AM users, manufacturers, educators, media and press, and researchers. This standard is not focused on supplying the standard safety procedures.
- **ASTM F2924-14:** This is another standard that particularly covers AM of biomedical titanium alloy (Ti-6Al-4V) components using electron beam and laser beam melting technologies (such as SLM, EBM, LENS®). The prime aim of this standard is to provide specifications for feedstock and supply chains for the production of biomedical devices, especially implants. The standard covers the following five main requisites:
  - It highlights that the AM Ti-6Al-4V components will be postprocessed through machining, grinding, electric discharge

Table 6.2 List of developed ASTM standards applicable to AM.

ASTM/ISO Designation	Title
Applicable to Design	
F3413-19	Guide for Additive Manufacturing—Design—Directed Energy Deposition
ISO/ASTM52915-16	Standard Specification for Additive Manufacturing File Format (AMF) Version 1.2
ISO/ASTM52910-18	Additive manufacturing—Design—Requirements, guidelines and recommendations
ISO/ASTM52911-1-19	Additive manufacturing—Design—Part 1: Laser-based powder bed fusion of metals
ISO/ASTM52911-2-19	Additive manufacturing—Design—Part 2: Laser-based powder bed fusion of polymers
Applicable to Materials and Processes	
F2924-14	Standard Specification for Additive Manufacturing Titanium-6 Aluminum-4 Vanadium with Powder Bed Fusion
F3001-14	Standard Specification for Additive Manufacturing Titanium-6 Aluminum-4 Vanadium ELI (Extra Low Interstitial) with Powder Bed Fusion
F3049-14	Standard Guide for Characterizing Properties of Metal Powders Used for Additive Manufacturing Processes
F3055-14a	Standard Specification for Additive Manufacturing Nickel Alloy (UNS N07718) with Powder Bed Fusion
F3056-14e1	Standard Specification for Additive Manufacturing Nickel Alloy (UNS N06625) with Powder Bed Fusion
F3091/F3091M-14	Standard Specification for Powder Bed Fusion of Plastic Materials
F3184-16	Standard Specification for Additive Manufacturing Stainless Steel Alloy (UNS S31603) with Powder Bed Fusion
F3187-16	Standard Guide for Directed Energy Deposition of Metals
F3213-17	Standard for Additive Manufacturing—Finished Part Properties–Standard Specification for Cobalt-28 Chromium-6 Molybdenum via Powder Bed Fusion
F3301-18a	Standard for Additive Manufacturing–Post Processing Methods–Standard Specification for Thermal Post-Processing Metal Parts Made Via Powder Bed Fusion

(*Continued*)

Table 6.2 (*Continued*)

ASTM/ISO Designation	Title
F3302-18	Standard for Additive Manufacturing—Finished Part Properties—Standard Specification for Titanium Alloys via Powder Bed Fusion
F3318-18	Standard for Additive Manufacturing—Finished Part Properties—Specification for AlSi10Mg with Powder Bed Fusion—Laser Beam
ISO/ASTM52901-16	Standard Guide for Additive Manufacturing—General Principles—Requirements for Purchased AM Parts
ISO/ASTM52904-19	Additive Manufacturing—Process Characteristics and Performance: Practice for Metal Powder Bed Fusion Process to Meet Critical Applications
ISO/ASTM52903-20	Additive manufacturing—Material extrusion-based additive manufacturing of plastic materials—Part 1: Feedstock materials
Applicable to Terminology	
ISO/ASTM52900-15	Standard Terminology for Additive Manufacturing— General Principles—Terminology
Applicable to Test Methods	
F2971-13	Standard Practice for Reporting Data for Test Specimens Prepared by Additive Manufacturing
F3122-14	Standard Guide for Evaluating Mechanical Properties of Metal Materials Made via Additive Manufacturing Processes
ISO/ASTM52902-19	Additive manufacturing—Test artifacts—Geometric capability assessment of additive manufacturing systems
ISO/ASTM52907-19	Additive manufacturing—Feedstock materials—Methods to characterize metallic powders
ISO/ASTM52921-13(2019)	Standard Terminology for Additive Manufacturing— Coordinate Systems and Test Methodologies

Source: https://www.astm.org/Standards/additive-manufacturing-technology-standards.html.

machining, polishing, or other likely processes in order to obtain the fine surface finish and critically controlled dimensional tolerances.

o   It helps the purchasers and manufacturers of Ti-6Al-4V components in defining their requirements and ensuring properties.

o   It recommends users to adopt this standard for obtaining components to meet minimum acceptance requirements.

o   It cites other resources to refer to while considering a stringent approach of purchasing.

o   It promotes the use of the Standard International (SI) unit system.

*   **ISO/ASTM52901-16:** This standard was developed in 2017 to offer detailed specifications asked by customers while purchasing AM parts, for example requirements of each part and specified qualities. The standard particularly emphasises on the following:

o   Pre-shipment inspection: This inspection is usually carried out by the supplier in order to ascertain that the part is delivered as per standard definitions.

o   Qualification part: Part fabricated prior to production process is used as a reference to verify the aspects of manufacturing processes and its characterization.

o   First production part: A part is produced with the same geometrical features, materials, and production facilities desired by the customer in order to verify to capability of machine tools and materials to satisfy the demand.

o   Reference part: A part with similar geometrical characteristics is put as a reference for characterization.

o   Acceptance: This defines the agreement between the manufacturer and customer that the delivered part is as per the order requirements.

o   Inspection plan: This includes a set of instructions specifying the verification of the manufacturing plan.

*   **ISO/ASTM 52910:** This presents a standard guideline for providing requirements and recommendations for using AM in product design. As the manufacturing of physical objects is achieved by the successive addition of materials, by following different procedural steps as discussed in previous chapters, it is important to define the key

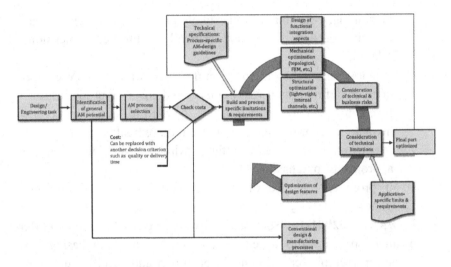

Figure 6.10   Design strategy for AM (adopted from ISO/ASTM 52910).

differences between the various categories of AM processes. This standard describes opportunities and design freedoms, discusses general design considerations, and lists 'red flags' for the designers to eliminate problems. This standard is particularly beneficial to newcomers to the AM industry. It provides key guidelines to the following type of users:

o   Designers who are designing a product to be fabricated through AM.
o   Students who are learning CAD and mechanical design.
o   Developers of AM design guidelines and design systems.

The design strategy suggested by ISO/ASTM 52910 is given in Figure 6.10.

6.4. Laboratory Protocol

This section provides a protocol to students for performing and estimating the various essential characteristics of an AM product. In particular, the suggested protocol is generally valid for fused deposition modelling products owing to its widespread use in educational institutions for students' learning. Following are the essential tools and equipment required for designing and developing test specimens: CAD software (SolidWorks® or Creo), FDM setup, software assistance (eg. Cura or Repetier), filament wire (acrylonitrile-butadiene styrene (ABS) or polylactic acid (PLA)), and SD Card or USB adapter. While operating the FDM system, the following safety practices must be followed:

o   Don't interrupt the printing operation and if it is required to stop the machine, press the stop button.
o   Don't put your hand or any tool in the printing area while the machine operates.
o   Don't touch the build platform as it will be hot.

Objective 1: To design, develop, and perform tensile test on FDM parts

Additional Equipment Required: Computerized Universal Testing Machine, Vernier Calliper, and test specimen

Learning Objectives: Understanding of stress-strain diagrams, tensile strength of a material, and significance of Young's modulus

Procedure:

1. Design the tensile test specimen as per ASTM-D638 standard by using CAD software.
2. Convert the CAD file into standard tessellation language (*.STL*) file format by using the CAD software's file export option.
3. Transfer the *.STL* file to the software assistance of the FDM setup.
4. Use ABS and PLA filament wires as the feedstock materials for building two different sets of tensile samples as per ASTM-D638 standard.
5. Print the tensile test specimens as per the factory recommended settings.
6. Note the start and end times of the printing process with the help of a digital stopwatch.

7. Upon completion, remove the as-built tensile sample from the build platform and the support structure by using soft hand tools.
8. Measure the breadth *b* and thickness *t* of the built sample halfway along the length.
9. Now calculate the original cross-sectional area of the sample *A* by using:

$$A = b \times t$$

10. Fix the sample in the jaws of the computerized tensile testing machine firmly.
11. Now start the machine and perform the test.
12. Store the stress vs. strain diagram obtained by the computer control software of the Universal Testing Machine. Calculate the stress *S* by using:

$$S = P/A$$

13. Using the available observations, fill in the following table.

Sample	Change in length (final length – original length)	Strain (change in length/ original length)	Load at Peak (N)	Stress at Peak (N/mm^2)	Load at Break (N)	Stress at Break (N/mm^2)
ABS						
PLA						

Objective 2: To design, develop, and perform compression test on FDM parts

Additional Equipment Required: Computerized Universal Testing Machine, Vernier Calliper, and test specimen

Learning Objectives: Understanding of stress-strain diagrams and compression strength of a material

Procedure:

1. Design the compression test specimen as per ASTM D695-10 standard by using CAD software.
2. Convert the CAD file into .*STL* and transfer to the software assistance of the FDM setup.
3. Use ABS and PLA filament wires as the feedstock materials for building two different sets of tensile samples as per ASTM D695-10 standard.
4. Print the compression test specimens as per the factory recommended settings.
5. Note the start and end times of the printing process with the help of a digital stopwatch.
6. Upon completion, remove the as-built compression sample from the build platform and the support structure by using soft hand tools.
7. Measure the original diameter (d_o) of the built sample halfway along the length.
8. Now calculate the original cross-sectional area of the sample A by using:

$$A = (\pi/4) \times d_o^2$$

9. Fix the sample in the jaws of the computerized compression testing machine firmly.
10. Now start the machine and perform the test.
11. Store the stress vs. strain diagram obtained by the computer control software of the Universal Testing Machine. Calculate the stress C by using:

$$C = P/A$$

12. Using the available observations, fill in the following table.

Sample	Change in diameter (final diameter – original diameter)	Strain (change in diameter/ original diameter)	Compressive Load at Break (N)	Compressive Stress at Break (N/mm^2)
ABS				
PLA				

Objective 3: To measure the linear and angular dimensions of FDM parts

Additional Equipment Required: Vernier Calliper, micro-meter, and test specimen

Learning Objectives: Understanding of change in linear and angular dimensions between a CAD model and fabricated FDM part

Procedure:

1. Design a cylindrical test specimen of diameter 30 mm and length 50 mm by using CAD software.
2. Convert the CAD file into *.STL* and transfer to the software assistance of the FDM setup.
3. Use ABS and PLA filament wires as the feedstock materials for building two different sets of test samples.
4. Print the test specimens as per the factory recommended settings.
5. Upon completion, remove the as-built compression sample from the build platform and the support structure by using soft hand tools.
6. Measure the final printed diameter d_f and final length l_f of the built sample by using micro-meter and Vernier Calliper, respectively.
7. Now calculate the change in diameter and length by using:

Change in diameter = ± (diameter of the CAD model – d_f)

Change in length = ± (length of the CAD model – l_f)

8. Using the available observations, fill in the following table.

Sample	Length of the CAD model (mm)	Diameter of the CAD model (mm)	Length of the printed sample (mm)	Diameter of the printed sample (mm)	Change in length (mm)	Change in diameter (mm)
ABS	50	30				
PLA	50	30				

Objective 4: To measure the surface roughness of FDM parts

Additional Equipment Required: Surface roughness tester and test specimen

> Learning Objectives: To understand the surface roughness of FDM parts
> Procedure:

1. Design a rectangular test specimen of length 50 mm, width 20 mm, and thickness 5 mm.
2. Convert the CAD file into *.STL* and transfer to the software assistance of the FDM setup.
3. Use the ABS filament wire as the feedstock materials for building two different sets of test samples.
4. Print the test specimens as per the factory recommended settings.
5. Upon completion, remove the as-built compression sample from the build platform and the support structure by using soft hand tools.
6. Measure the surface roughness R_a of the printed sample at five different locations and calculate the average value.
7. Using the available observations, fill in the following table.

Sample	Roughness (µm) #1	Roughness (µm) #2	Roughness (µm) #3	Roughness (µm) #4	Roughness (µm) #5	Average roughness (µm)
ABS						
PLA						

Glossary of Key Terminologies

ABS: Acrylonitrile-Butadiene Styrene
AM: Additive Manufacturing
ASM: American Society for Metals
ASTM: American Society for Testing and Materials
BHN: Brinell Hardness Number
CAD: Computer-Aided Design

CEN: European Committee for Standardization
EBM: Electron Beam Melting
FDM: Fused Deposition Modelling
ISO: International Organization for Standardization
LENS®:   Laser Engineered Net Shaping
NDT: Non-Destructive Testing
PLA: Polylactic Acid
RPT: Rapid Prototyping and Tooling
SLM: Selective Laser Melting
.STL: Standard Tessellation Language

Exercise Questions

Q1: What is metrology? What are the different types of tools used in metrology?

Q2: What is testing? Differentiate between destructive and non-destructive testing.

Q3: Discuss any three different types of non-destructive testing techniques.

Q4: What are the key factors impacting measurement?

Q5: What are (a) tensile test, (b) hardness test, and (c) impact test?

Q6: Give the full names of ISO, ASTM, and CEN.

Q7: Explain the standard development process in brief by using a suitable diagram.

Q8: Explain ISO/ASTM52901-16 standard for AM in short.

Q9: What was the main purpose of developing the ASTM F-42 committee?

Q10: With the help of a suitable diagram, discuss the stress vs. strain diagram.

Q11: Explain ASTM F2924-14 standard in brief.

References

1. Willcox M, Downes G. A brief description of NDT techniques. Toronto: NDT Equipment Limited. 2003.
2. Lu QY, Wong CH. Applications of non-destructive testing techniques for post-process control of additively manufactured parts. *Virtual and Physical Prototyping*. 2017;12(4):301–321.

3. https://www.asminternational.org/documents/10192/3465262/05105G_Chapter_1.pdf/e13396e8-a327-490a-a414-9bd1d2bc2bb8.

4. Hamad AJ. Size and shape effect of specimen on the compressive strength of HPLWFC reinforced with glass fibres. *Journal of King Saud University-Engineering Sciences*. 2017;29(4):373–380.

5. https://www.asminternational.org/documents/10192/1849770/06156G_Sample.pdf4.

6. Supe JD, Gupta MK. Flexural Strength — A measure to control quality of rigid concrete pavements. *International Journal of Scientific & Engineering Research*. 2014;5(11):46–57.

7. https://www.asminternational.org/documents/10192/1849770/06671g-ch.pdf.

8. https://aybu.edu.tr/muhendislik/makina/contents/files/HARDNESS%20TEST(1).pdf.

9. Makes A, Collaborative AA. Standardization roadmap for additive manufacturing. (February), Public Draft.

10. Bourell DL, Leu MC, Rosen DW. Roadmap for additive manufacturing: identifying the future of freeform processing. The University of Texas at Austin, Austin, TX. 2009.

11. Campbell I, Diegel O, Kowen J, Wohlers T. Wohlers report 2018: 3D printing and additive manufacturing state of the industry: Annual worldwide progress report. Wohlers Associates; 2018.

12. Monzón MD, Ortega Z, Martínez A, Ortega F. Standardization in additive manufacturing: Activities carried out by international organizations and projects. *The International Journal of Advanced Manufacturing Technology*. 2015;76(5–8):1111–1121.

13. Gupta N, Weber C, Newsome S. Additive manufacturing: Status and opportunities. Science and Technology Policy Institute, Washington. 2012.

Index

Printed in the United States
By Bookmasters